NANOTECHNOLOGY SCIENCE AND TECHNOLOGY

NANOSHEETS AND NANOSPHERES

TYPES, APPLICATIONS AND RESEARCH INSIGHTS

NANOTECHNOLOGY SCIENCE AND TECHNOLOGY

Additional books in this series can be found on Nova's website under the Series tab.

Additional e-books in this series can be found on Nova's website under the eBook tab.

NANOTECHNOLOGY SCIENCE AND TECHNOLOGY

NANOSHEETS AND NANOSPHERES

TYPES, APPLICATIONS AND RESEARCH INSIGHTS

DEBORAH WRIGHT
EDITOR

nova publishers
New York

Library of Congress Cataloging-in-Publication Data

ISBN: 978-1-53610-203-1

Published by Nova Science Publishers, Inc. † New York

CONTENTS

Preface		**vii**
Chapter 1	Graphene-Based Nanosheet Functionalized Membranes – Industrial Applications *Jenny Lawler*	**1**
Chapter 2	Transition Metal Dichalcogenide Nanosheets for Sensor Platforms *Yong Wang and Qianfen Zhuang*	**41**
Chapter 3	Fabrication and Function of "Polymer Nanosphere Multilayered Organization" *Manami Iizuka and Atsuhiro Fujimori*	**85**
Chapter 4	Oxygen Electrochemistry on Two Dimensional Nanosheets *Hengcong Tao and Zhenyu Sun*	**133**
Index		**171**

PREFACE

This book discusses new research on nanosheets and nanospheres. Chapter One provides a review of the incorporation of, in particular, carbon nanosheets such as graphene and derivatives such as graphene oxide into polymeric and ceramic membranes for applications in a wide variety of industries. Chapter Two summarizes some representative works of applications of transition metal dichalcogenide (TMD) nanosheets in sensor platforms. Chapter Three introduces on a new functionality exhibited by "polymer nanosphere multilayered organization," a new type of molecular organization, and the relationship between their structure and function. Chapter Four summarizes recent advances in the electrochemical oxygen-relative reactions, such as oxygen reduction reaction and oxygen evolution reaction by using electrocatalysts of graphene and its derivatives and also other 2D materials.

Chapter 1 – Membrane separations is the umbrella term typically given to operations including microfiltration (MF), ultrafiltration (UF), nanofiltration (NF), reverse osmosis (RO), forward osmosis (FO), membrane distillation (MD) and pervaporation (PV). These technologies are widely employed in a diverse number of industries. The primary trade-off that researchers strive to enhance, regardless of the application (whether gas or solvent separations, water purification or ion/proton transport) is the balance between the flux and the selectivity (e.g., retention of valuable proteins, rejection of unwanted compounds or organic matter and micro-organisms, diffusion of methanol in fuel cells). Foulant molecules such as bacteria, proteins or organic matter can lead to a decline in membrane performance, flux and selectivity. Foulants can adhere or adsorb on the membrane surface and/or in the membrane pores, requiring process controls such as backflushing for reversible fouling, and requiring chemical cleaning in the case of irreversible fouling. This increases the operating and maintenance costs associated with the process and causes a decline in the lifespan of membranes. As such, much research has focused on methods for the enhancement of the hydrophilicity, permeability and antifouling ability of membranes. A number of different approaches have been investigated to this end incorporating the use of graphene based nanosheets such as GO (graphene oxide) and rGO (reduced GO), including, for example, the blending of nanosheets as fillers within polymeric membranes, the incorporation of nanosheets within the thin film active layer of RO/NF membranes, and many more; some of these approaches are summarized within this chapter. This chapter provides a review of the incorporation of, in particular, carbon nanosheets such as graphene and derivatives such as GO, into polymeric and ceramic membranes for application in a wide variety of industries. The focus is on the enhancement of productivity and reduction in fouling characteristics provided by appropriate incorporation of nanosheets during membrane fabrication. Case studies detailing the industries/processes for which they are employed are summarized within each section, the challenges associated with the particular operations and how the incorporation of nanosheets can help to address those challenges are discussed, along with directions for future research.

Chapter 2 – In the past few years, transition metal dichalcogenide (TMD) nanosheets have aroused tremendous attention within the scientific community due to a wealth of unique new physics and chemistry that emerges as its thickness goes down to single or few layer. The unusual and fascinating physicochemical properties of these TMD nanosheets make them intriguing candidates for use in optoelectronic devices, energy storage devices and catalysis. However, in very recent times, there was a significant emerging trend in their utilization in sensors. Therefore, in this book chapter, the authors attach great importance to this new trend for the sensing applications, and summarize some representative works of applications of TMD nanosheets in sensing area. Moreover, current challenges and future perspectives in the sensing area are also outlined.

Chapter 3 – This chapter introduces on a new functionality exhibited by "polymer nanosphere multilayered organization," a new type of molecular organization, and the relationship between their structure and function. The polymer nanosphere multilayered organization is a fine structural material formed by the accumulation of single-particle layers of a hydrophobic polymer at the air/water interface; these single-particle layers have uniform height along the *c*-axis. The authors constructed a multiparticle layered organization of aromatic polyamides with rigid main chains and flexible side chains by the Langmuir–Blodgett (LB) technique, which resulted in a highly regular arrangement along the *c*-axis. The particle arrangement was estimated by performing out-of-plane X-ray diffraction (XRD) analysis and atomic force microscopic (AFM) observation. The results suggest that a double-particle layered structure (Y-type) is formed by the LB technique, forming amphiphilic particles at the air/water interface. Copolymers with highly hydrophobic carbazole contents and both hydrogenated and fluorinated side-chains also formed a single-particle layer at the air/water interface and exhibited multiparticle layers by a LB technique. Therefore, it is possible to control the formation of single- and double-particle layered structure using these techniques. Further, it was found that multiparticle layered organization of polymer nanospheres and polymer nanosheets could be formed simultaneously with the same component material. By employing the "alternate compression−relaxation method," high-density, low defect particle layers are formed with a clear increase in their crystallite sizes. In the case of a ternary comb copolymer containing a carbazole ring, one particle is formed by the assembly of approximately 60 units of collapsed monolayer-like double layers. This structure is stabilized by the formation of side-chain crystals in the interlayer, with oriented π–π stacking of carbazole rings, resulting in enhanced fluorescence emission intensity.

Chapter 4 – Two-dimensional (2D) nanosheets have drawn great attention because of their interesting electronic and chemical properties since the isolation of graphene in 2004. This book chapter would summarize recent advances in the electrochemical oxygen-relative reactions, such as oxygen reduction reaction and oxygen evolution reaction by using electrocatalysts of graphene and its derivatives and also other 2D materials. From the mechanistic point of view, the link between structure and activity of these 2D nanosheets is discussed. Strategies that are effective to adjust electronic structure in these 2D materials and thereby catalytic activities are emphasized. Future opportunities and challenges in this field are outlined.

In: Nanosheets and Nanospheres
Editor: Deborah Wright
ISBN: 978-1-53610-203-1

Chapter 1

GRAPHENE-BASED NANOSHEET FUNCTIONALIZED MEMBRANES – INDUSTRIAL APPLICATIONS

Jenny Lawler**, PhD
School of Biotechnology and DCU Water Institute,
Dubln City University, Dublin, Ireland

ABSTRACT

Membrane separations is the umbrella term typically given to operations including microfiltration (MF), ultrafiltration (UF), nanofiltration (NF), reverse osmosis (RO), forward osmosis (FO), membrane distillation (MD) and pervaporation (PV). These technologies are widely employed in a diverse number of industries. The primary trade-off that researchers strive to enhance, regardless of the application (whether gas or solvent separations, water purification or ion/proton transport) is the balance between the flux and the selectivity (e.g., retention of valuable proteins, rejection of unwanted compounds or organic matter and micro-organisms, diffusion of methanol in fuel cells).

Foulant molecules such as bacteria, proteins or organic matter can lead to a decline in membrane performance, flux and selectivity. Foulants can adhere or adsorb on the membrane surface and/or in the membrane pores, requiring process controls such as backflushing for reversible fouling, and requiring chemical cleaning in the case of irreversible fouling. This

* Corresponding Author address; Email:jenny.lawler@dcu.ie.

increases the operating and maintenance costs associated with the process and causes a decline in the lifespan of membranes. As such, much research has focused on methods for the enhancement of the hydrophilicity, permeability and antifouling ability of membranes.

A number of different approaches have been investigated to this end incorporating the use of graphene based nanosheets such as GO (graphene oxide) and rGO (reduced GO), including, for example, the blending of nanosheets as fillers within polymeric membranes, the incorporation of nanosheets within the thin film active layer of RO/NF membranes, and many more; some of these approaches are summarized within this chapter. This chapter provides a review of the incorporation of, in particular, carbon nanosheets such as graphene and derivatives such as GO, into polymeric and ceramic membranes for application in a wide variety of industries. The focus is on the enhancement of productivity and reduction in fouling characteristics provided by appropriate incorporation of nanosheets during membrane fabrication. Case studies detailing the industries/processes for which they are employed are summarized within each section, the challenges associated with the particular operations and how the incorporation of nanosheets can help to address those challenges are discussed, along with directions for future research.

Keywords: membranes, graphene, graphene oxide, desalination, pharmaceuticals, fuel cells

INTRODUCTION

Graphene is a two-dimensional single-layered nanomaterial arranged in a hexagonal honeycomb lattice structure, consisting of sp^2-hybridized carbon atoms (Figure 1). The potential for graphene in membrane separations was first elucidated by Nair et al. [1], when they discovered that thin graphene oxide (GO) membranes allowed transport of water only but did not allow passage of gases or other liquids, attributed to the low-friction flow of a monolayer of water through capillaries formed by closely spaced graphene sheets. The spacing between nanosheets is typically of the order of 0.3 – 0.7 nm, which is ideal for the permeation of water while blocking the transport of larger molecules. The capillaries would be narrowed or blocked by water molecules, consequently not allowing permeation of other molecules [1]. Carbon nanosheets such as graphene and derivatives such as graphene oxide (GO) and reduced graphene oxide (rGO) have received increasing attention in the membranes literature due to their interesting properties and potential for enhancing the performance of polymeric or ceramic membranes by fabrication of nanocomposites or nanohybrids. Research ranges from pure graphene films and graphene in which nanopores have been introduced, to GO and rGO, in which nanoscale defects naturally occur during the manufacture process. Mass transport is generally governed by steric exclusion, electrostatic interaction, and chemical binding in the case of GO and rGO [2]. The resulting membranes, a combination of polymers or ceramics and nanomaterials, are increasingly being seen as a solution to, in particular, environmental problems such as the challenge of providing clean water free from contaminants or clean fuel. GO in particular, has an ultra-thin two dimensional structure and has been shown to possess many oxygen containing functional groups (e.g., carboxylic acid groups at edges, and phenol hydroxyl and epoxide groups at the basal planes), which can confer hydrophilic properties on the membrane [3], and provide reactive sites for membrane functionalization, thus displaying excellent water permeability (Figure 1). GO dispersions typically display a brown colour while reduction to rGO darkens the colour towards black. Spacing between nanosheets leaves interconnected channels through which water molecules can flow, and a convoluted tortuous path is generally accepted to follow mainly over the hydrophobic non-oxidized surfaces rather than the oxidized, hydrophilic region [4, 5]. Nanosheets provide the opportunity to produce tailored membranes by allowing manipulation of membrane structural properties, thermal and mechanical stability, hydrophilicity, charge and charge density, and porosity and roughness; they have the potential to enhance the balance between permeability and selectivity while minimising the fouling that can be the major limitation of membrane separations [6, 7]. While much research has focussed on solvent separation and gas separation, these topics are not dealt with in this chapter. Here, case studies are presented on the preparation of membranes incorporating carbon based nanosheets for the enhancement of membrane properties over their non-nanosheet containing ceramic or polymeric counterparts.

Classification of Membrane Separations

Membrane separations is the umbrella term typically given to operations including microfiltration (MF), ultrafiltration (UF), nanofiltration (NF), reverse osmosis (RO) (Table 1), forward osmosis (FO), membrane distillation (MD) and pervaporation (PV). These technologies are widely employed in a diverse number of industries, for example the purification of proteins in the pharmaceutical industry (UF), the treatment of wastewater (MF and UF) and drinking water (MF, UF, NF, RO), the desalination of seawater and brackish water (NF, RO, FO, PV), cleaning of blood during haemodialysis (UF), solvent separation (PV) and many more. Two primary, common goals in all settings is enhanced productivity (i.e., the flux of treated permeate) and selectivity (e.g., retention of valuable proteins, rejection of unwanted compounds or organic matter and micro-organisms). A complete introduction to membranes and the physics underlying these different and distinct separation systems can be found in a number of key publications and textbooks [8–11], and the author assumes here some prior knowledge of membrane separation processes.

MEMBRANE PREPARATION

Graphene derivatives such as GO are attractive due to their ease and low cost of manufacture in comparison to pristine graphene. GO is typically produced from an aqueous dispersion of graphite oxide, followed by exfoliation in a solvent, typically using a Hummer's or modified Hummer's and Offeman method [12, 13]. Figure 2 a) shows clearly the thinness of the GO nanosheet, while the wrinkled surface is clearly visible in a) and b). The pore size/nanosheets spacing is adjustable, by tuning the properties of the GO nanosheets including by inclusion of filler molecules such as TiO_2 or other nanoparticles [4, 14–16].

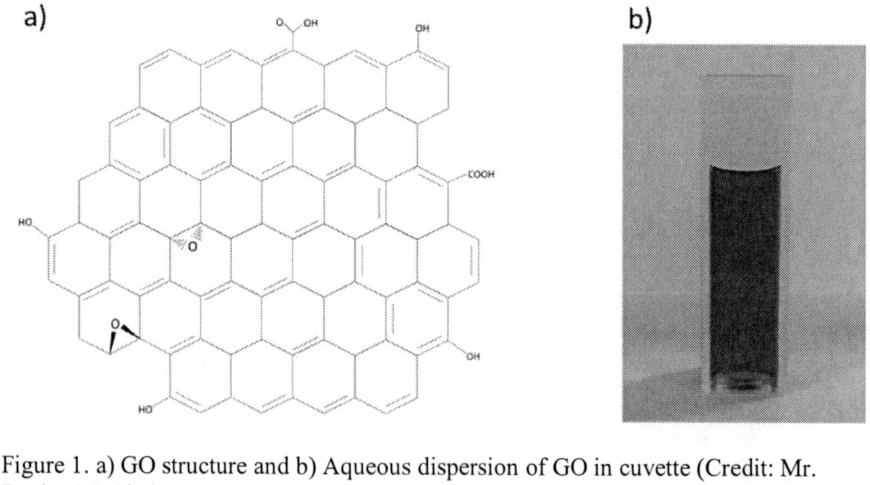

Figure 1. a) GO structure and b) Aqueous dispersion of GO in cuvette (Credit: Mr. Declan McGlade).

Table 1. Typical Membrane Filtration Processes

Technique	Size Classification	Smallest Type of Molecule Typically Rejected	Membrane Type	Pressure Range (bar)	Applications
MF	0.1 – 10 μm	Bacteria	Porous	0.5 - 3	Separation of cells/bacteria from liquid
UF	1 – 100 nm	Viruses	Microporous	1-10	Separation of proteins
NF	0.5 – 5 nm	Divalent Ions	Microporous / Dense	7-40	Separation of dyes, water softening
RO	<1 nm	Monovalent Ions	Dense	25-100	Desalination of sea and brackish water

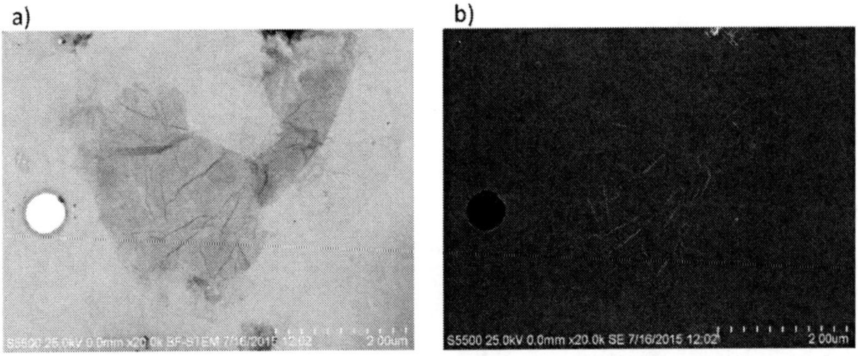

Figure 2. FESEM Image of GO in a) Transmission and b) Secondary Electron modes (Credit: Mr. Declan McGlade).

Fabrication of many nanocomposite membranes is based on phase inversion (e.g., the non-solvent induced phase separation (NIPS) method or thermally induced phase separation (TIPS) method), in which nanosheets are dispersed in polymer solutions prior to the casting of porous or microporous membranes, and can be prepared in either flat sheet or hollow fiber configurations. During the NIPS process, the hydrophilic nature of GO sheets means that the GO converges to the membrane and pore surfaces [17] during precipitation in aqueous solution. This type of membrane is mainly used in MF or UF processes due to its typically porous structure, and the structure most often consists of a thin dense skin layer over a more macroporous sublayer (asymmetric structure), formed due to the slower precipitation of the polymer below the skin layer [7]. More dense membranes, such as NF or RO membranes, are typically of a thin film composite (TFC), multi-layer type structure, composed of a thin dense polymer film or barrier layer on top of one or more porous support layers [18] (Figure 3). The porous membranes previously outlined often form the porous support layer for TFC membranes, and carbon nanosheets can be incorporated into either the support layer or the thin barrier layer. GO membranes have been prepared by layer-by-layer (LbL) deposition of GO suspension or else simple vaccum filtration onto a support layer [19], while membranes have also been prepared by electrospinning, spin coating or drop casting [20, 21]. While vacuum/solution filtration for deposition of pure GO membrane layers appears attractive, the mechanical stability under hydrated conditions can be insufficient, with GO layers tending to slough off the support layer under typical crossflow/pressurized conditions. In an aqueous environment, the layers become negatively charged, leading to electrostatic repulsion that can exceed the van der Waals or hydrogen bonding forces that hold the GO layers together [14, 22]. As such, surface layer deposited GO must generally be crosslinked to maintain membrane integrity [19, 23], and this is often an approach taken for immobilization of GO on the surface of ceramic membranes.

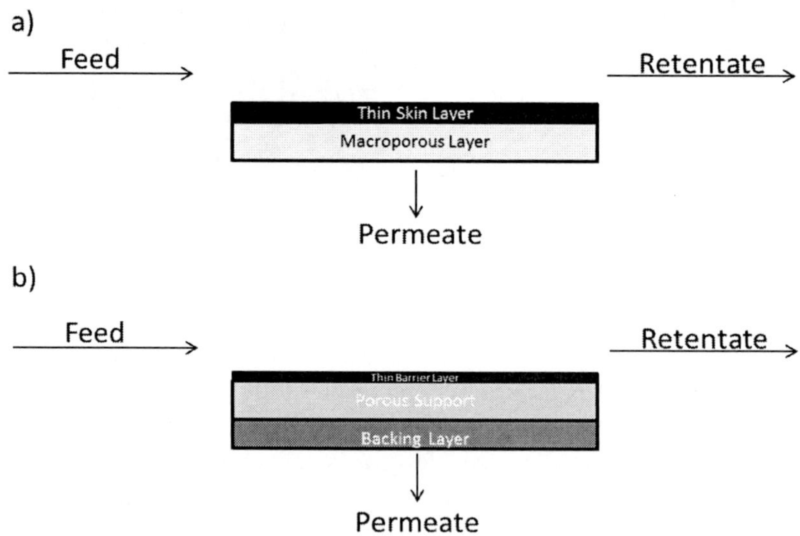

Figure 3. Membrane Structure a) Asymmetric macroporous membrane b) Thin film composite membrane.

CARBON NANOSHEET MEMBRANES FOR DESALINATION

Fresh water scarcity is an increasing problem globally, which has led to a strong drive for research in the area of water reclamation and seawater desalination. The need for fresh water both domestically and industrially (agriculture, food production etc.) has increased at unprecedented rates in recent years. The most common type of membrane for water desalination is reverse osmosis (RO), as it rejects mono- and divalent salts from sea and brackish water, while energy requirements for RO have decreased to levels that are more desirable than thermal methods [24]. Reverse osmosis utilises two opposing forces in effecting separation between two phases, the concentration gradient and the pressure gradient. During osmosis, flow of pure water through the membrane to the salt rich side will occur. During reverse osmosis, high pressure is applied on the salt rich side of the membrane, counteracting the osmosis effect of the concentration gradient.

In recent years, alternative membrane processes such as forward osmosis (FO) and nanofiltration (NF) have garnered increasing attention in the literature and industrially, as they have some potential advantages including lower energy requirements.

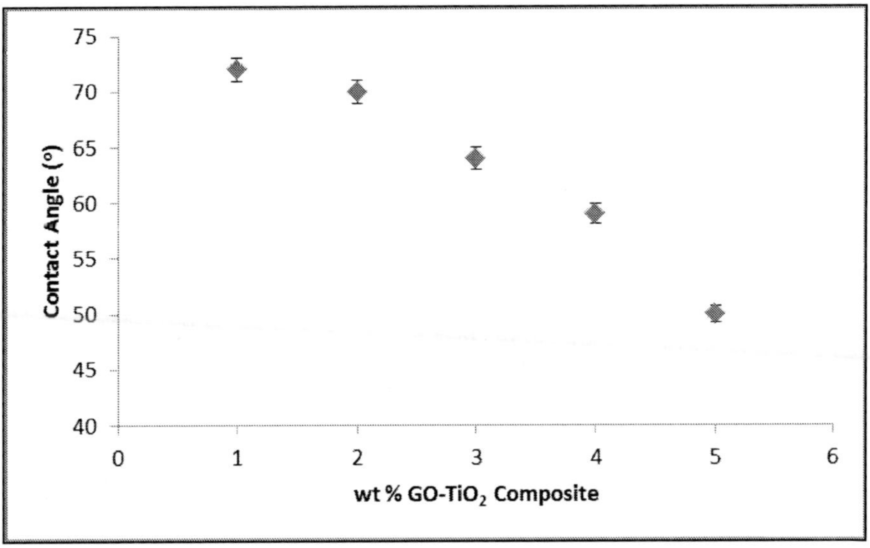

Figure 4. Increasing hydrophilicity with increasing GO-TiO_2 content in PSf hybrid membranes.

Nanoporous graphene has proven to be an excellent candidate for desalination, with membranes consisting of nanopores created in single layer graphene sheets using an oxygen plasma etching process displaying almost 100% salt rejection with extremely fast fluxes of 6 x 10^5 L/m^2/hr (LMH) [25] and similar membranes prepared using double layer graphene sheets etched using selective ion beam etching also showing great promise [26], under pressure driven conditions.

While graphene itself would be an excellent candidate for removal of salts while offering fast water transport, due to difficulties in large scale manufacture and stability as previously discussed, GO or rGO are the most attractive options currently for membrane preparation/modification.

The most commonly employed membranes for desalination have been polyamide thin film composite (PA-TFC) membranes, due to their good mechanical stability and excellent separation capabilities. PA-TFC membranes typically consist of a non-woven fabric support layer of approx. 120 μm thickness, topped with a sublayer of polymer (often polyethersulfone (PES) or polysulfone (PSf) with a thickness of approximately 40 μm, on which an interfacial polymerization is performed to fabricate a surface active polyamide layer of approximately 200 nm (Figure 3). The surface active PA layer is typically the part of the membrane that governs the separation process. This type of membrane has superceded the previously employed cellulose acetate (CA) membranes, as they can tolerate larger ranges of operating pH, temperature and pressure. Although improving, energy consumption and membrane fouling remain a problem in RO membranes, and this is an area of much research. Fouling mitigation measures such as enhancing the hydrophilicity and decreasing the surface roughness of the membrane is a common approach for improving membrane performance (Figure 4). Chlorine resistance is another pressing issue – chlorine residues that typically remain in pre-treated water can be damaging to the membrane surface active layer. Mitigation of biofouling via the membrane itself is something that has potential – with incorporation of antiomicrobial compounds such as copper or silver in hybrid nanocomposite with GO or rGO showing synergistic effects in water disinfection [27, 28].

Modification and functionalization of both surface active layers and membrane sublayers have been investigated, and the incorporation of nanostructures such as GO shows great promise due to fast water channels, high permeability and excellent hydrophilicity [29, 30]. It is possible to modify the membrane properties either by surface modification (such as coating or grafting) or by modification of the composition of the sublayer or active layer, for example using amide coupling between carboxyl groups of graphene oxide and carboxyl groups of the polyamide active layer [31]; in addition the mechanical properties of the membrane can be enhanced for example by crosslinking with materials such as chitosan [32].

LbL deposition of GO multilayers coated on a PA-TFC RO membrane displayed an increase in surface hydrophilicity and decrease in surface roughness, leading to improved antifouling characteristics. The GO layer acted as a protection against chlorine damage to the PA layer, so cleaning or chlorine exposure did not result in degradation of salt rejection [33]. The LbL deposition of negatively charged graphene oxide (GO) nanosheets on a porous poly(acrylonitrile) support yielded a tight nanoporous membrane with MWCO of <350, ideal for multivalent ion desalination / water softening in FO process [34]. 0.02 wt% rGO-TiO_2 nanocomposite embedded in the PA layer of TFC RO membranes yielded significant improvements in chlorine resistance, antifouling ability, hydrophilicity and enhanced surface roughness, with fluxes of >51 LMH and good salt rejection [23], while GO-TFC membranes produced by incorporation of GO in the PA layer yielded poorer salt rejection (~93% NaCl rejection) but better flux (59 LMH) [6]. Membranes prepared via GO-monolayer titania with UV reduction from GO to rGO demonstrated excellent ion rejection (<5% permeation) with good water permeabilities provided by photoinduced hydrophilic conversion of the titania [35]. Incorporation of GO into a PSf support layer with subsequent interfacial polymerization to form PA barrier layer yielded membranes suitable for operation in either RO mode (with salt rejection of almost 99% and flux of almost 2 LMH per bar applied transmembrane pressure), or FO mode [36].

NF membranes sit somewhere in between UF and RO, with "tight" NF membranes being compared to "loose" RO membranes, and "loose" NF membranes being akin to "tight" UF membranes. Looser NF membranes are of significant interest for water softening, for example with layer-by-layer assembled polyethyleneimine-GO membranes exhibiting a positively charged surface for salt rejection [37]. Tight NF membranes have a salt rejection for NA^+ of L 90%, and typically have a MWCO of < 200 Da. Many parameters of the membrane are implicated in the control of solute rejection and flux, including the operating conditions, membrane fabrication technique, composition of the active layer (and sublayer), and presence and concentration of active layer cross-linking agent. Transport of solutes across the membrane is similar to RO membranes in terms of diffusional limitation and steric exclusion. The solute rejection depends strongly on the solute type, which includes charge valency, diffusion coefficient, and hydration energy [38]. The flux in NF membranes tends to decrease with increasing salt concentration e.g., NaCl or Na_2SO_4 in the feed solution, due to the increase in feed osmotic pressure [39] or decrease in GO zeta potential with increase in NaCl concentration in lamellar GO membranes [40], while decrease in surface roughness tends to improve filtration characteristics by increasing the antifouling properties of the membrane. Solute rejection was shown to be improved when chemical functional groups (such as sulfonic or carboxylic acid groups) were incorporated into NF membranes [41], while allowing for a decrease in pressure requirements and allowing for thinner membranes with a concomitant increase in permeate flux, while in addition, an increase in hydrophilicity of membranes was shown to enhance their antifouling capabilities [42], suggesting that GO would be an excellent target for research into NF membrane modification. In addition, molecular dynamic simulations have shown that graphene layers with nanopore diameters could offer high salt rejection and excellent fluxes, far improving on what is currently possible with RO membranes [43–45], supported by recent experimental work [25, 26].

Vacuum deposition of GO based films onto support membranes has been used to prepare tight NF membranes (Figure 5). Ultrathin (<60 nm thick) rGO layer on polymeric MF membranes provided good salt rejection (20 – 60% of ions) [46], while better salt rejection but lower flux were found for membranes with MWCNTs incorporated into the rGO nanolayer when deposited on Al_2O_3 MF membranes [47]. The antifouling nature of the membranes to humic acid (HA) and sodium alginate was good but protein fouling was not ideal. Improvements in flux due to enhancement in steric effects were realized by better dispersion and mixing of the CNT/GO layer by addition of pluronic surfactants (block copolymers) which were vacuum filtration deposited onto anodic Al_2O_3 MF membranes [48]. Again, drying related shrinkage of films of this type have been reported and damage to films is possible under crossflow and high TMP conditions, although efforts to stabilise the film showed reasonable success by incorporation of a sacrificial layer of PMMA with in-situ reduction of GO using UV [49]. Dip coating of negatively charged GO film onto positively charged polyamide-imide/polyethylene-imine hollow fibre membranes yielded a significant flux enhancement without sacrifice of salt rejection, with good mechanical stability of dip coated layer [50].

Interfacial polymerization of GO/rGO/polymer mixtures on support layers has also been shown to be effective in preparation of NF membranes. Excellent salt rejection of 88% for NaCl and 97% for $MgSO_4$, with the latter enhanced by electrostatic repulsion due to the membrane charge, were achieved using membranes prepared by interfacial polymerization of mixtures of rGO and polyamide (PA) on PSf UF membrane supports [51]. The addition of rGO was shown to enhance the membrane antifouling properties over pure PA membranes when tested with HA and bovine serum albumin (BSA). rGO-TiO_2 has been included as a hydrophilic modifier in the active layer of TFC membranes, leading to a decrease in surface roughness and increase in the fouling resistance of the membranes. Fluxes of ~ 55-60 LMH were achieved, with > 90% rejection of the multivalent SO_4^{2-} anion and 35% of monovalent chloride anion [23]. Increase in the rGO-TiO_2 content of the membrane above 0.02 wt% led to a decrease in the salt rejection, potentially due to increase in defects in the active layer. Ring opening polymerization reaction was used to stably immobilize epoxy groups of GO to amino groups of an active layer of O-(carboxymethyl)-chitosan on a PSf membrane support. The GO nanosheets were immobilized on the surface, and again increase in negative charge was seen with rejections of > 97% for Na_2SO_4 and >60% for NaCl, with fluxes of > 20 LMH [52]. Loose NF membranes with zwitterionic GO grafted with poly(sulfobetaine methacrylate) brushes (GO-PSBMA) were prepared by phase inversion of mixtures of GO-PSBMA and PES [53]. The zwitterionic nature of the membranes allowed for reduction in fouling potential, and the loose nature of the membranes allowed retention of dyes but permeation of salts [54]. Polyethylene glycol modified graphene oxide (PEG/GO) nanosheets were incorporated into the PA layer, with UV irradiation shown to enhance permeate flux while maintaining rejection of NaCl and Na_2SO_4 of 63.5% and 93% respectively [31]. The flux enhancement was attributed to the standard increase in hydrophilicity due to inclusion of GO in the structure, with UV irradiation leading to longer hydrophilic PEG backbone in the structure of the thin layer.

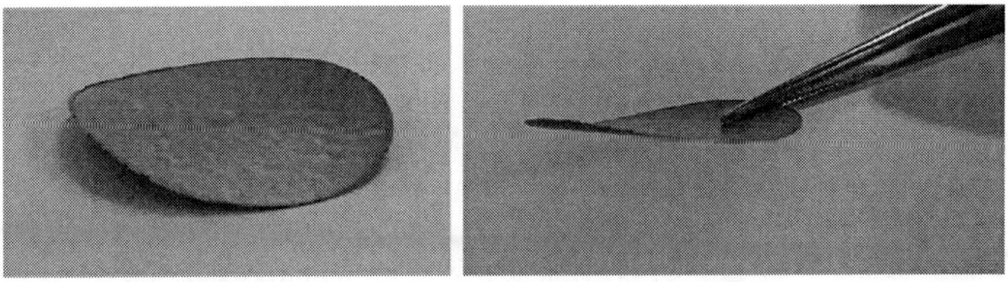

Figure 5. Freestanding rGO films (Credit: Mr. Declan McGlade).

While reverse osmosis is the most commonly used membrane based desalination method, and NF is gaining increased interest particularly for water softening, pervaporation (PV) has attracted attention as an alternative approach in recent years as it offers lower energy consumption and ease of operation, not requiring the high pressures of RO/NF separation [19]. Pervaporation utilises a dense membrane (polymeric, inorganic or hybrid) through which a specific component may preferentially permeate and evaporate. The feed is typically heated to approx. 45-60 °C, allowing the use of low-grade waste or solar energy. A gradient of concentration and vapour pressure drives the mass transfer. A low vapour pressure on the permeate side is generated by either vacuum, an air gap, or a sweep flow, and it is generally accepted that the separation is effected under solution-diffusion mechanism [55]. While pervaporation has mostly been utilized for solvent dehydration and separation of organic mixtures, it demonstrates high salt rejection (above 99%) due to the non-volatile nature of the salts. The impact of feed concentration on permeability is less in PV than in RO, as the high osmotic pressures that have to be overcome in RO do not apply in PV due to the phase change nature of the mass transfer (rather than pressure driven) – this means that PV is well suited to handle high salinity feeds, and can be more fouling resistant than RO. PV membranes typically do not suffer with fouling from adsorption or pore plugging, and membrane wetting and salt leakage can be avoided.

The flux in PV membranes remains low, typically less than 6-10 LMH, and is strongly dependent on the physicochemical characteristics of the membrane material. As such significant interest lies in the exploration of highly hydrophilic materials (such as GO) with open passages for water transport, for high flux while maintaining salt rejection.

Membranes utilizing a 3-dimensional GO framework supported on Al_2O_3 tubular membranes by covalent synthesis and vacuum filtration have been prepared, using crosslinking agents to prepare the 3-D framework, allowing separation of the GO layers, enhancing mechanical stability while allowing flux and rejection of salts. Salt rejection of > 99.5% has been achieved with fluxes > 11 LMH using 1, 4-phenylene diisocyanate (PDI) crosslinker [19] and > 48 LMH using polydopamine (PDA) [56].

In a process similar to pervaporation, membranes suitable for direct contact membrane distillation were prepared by immobilization of GO on the surface of a hydrophobic polytetrafluoroethylene (PTFE) membrane, achieving fluxes of up to 97 LMH [57].

Photocatalytic Carbon Nanosheet Membranes

In order to address the issue of membrane fouling discussed previously, efforts have been made to incorporate photocatalytic elements into membranes as foulant degradation can be carried out in-situ, hence reducing the requirement for chemical cleaning or membrane replacement. Photocatalyst compounds can be incorporated within the membrane matrix or deposited or bonded onto the membrane surface, depending on the application.

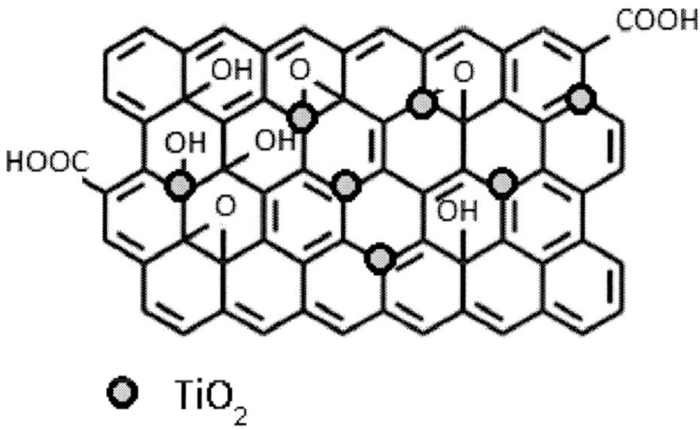

Figure 6. GO-TiO_2 nanocomposite prepared via sol-gel process.

A graphitic carbon nitride nanosheet/reduced graphene oxide/cellulose acetate composite photocatalytic membrane (g-C_3N_4 NS/rGO/CA) was prepared by vacuum filtration onto the surface of commercial cellulose acetate (CA) microfiltration membrane, with increase in flux over CA membrane alone. While the inclusion of rGO within the membrane matrix enhanced the antifouling properties due to increase in hydrophilicity, visible light irradiation further enhanced the flux owing to degradation of organic compounds responsible for membrane fouling [37]. Titanium dioxide (TiO_2) is one of the most common photocatalysts employed, and GO has been shown to enhance the photocatalytic activity of immobilized TiO_2 as GO sheets can facilitate separation of photo-induced electron-hole, suppressing charge recombination [58]. GO-TiO_2 can be prepared via simple mixing or via sol-gel process (Figure 6), to avoid agglomeration of TiO_2 nanoparticles at higher concentrations (Kumar et al. 2016).

GO/TiO_2/Fe_2O_3 was deposited via vacuum filtration onto glass fibre filters, exhibiting enhanced adsorption of HA with photocatalytic degradation under simulated solar irradiation under constant flux of 145 LMH, with 40% of the pressure drop associated with the same membrane under dark conditions [59]. GO-TiO_2 photocatalytic membranes have been prepared by LbL deposition onto PSf support [60], as a microspherical hierarchy onto CA membrane [61], immobilized on alginate hollow fibres via a wet-dry spinning process [62], decorated onto poly(vinylidene difluoride-co-trifluoroethylene) (P(VDF-TrFE)) copolymer fibres and electrospun to prepare nanofibrous mats [63] and immobilized by dip-coating onto ceramic UF and NF monoliths [64]. In all cases increase in membrane flux due to enhanced hydrophilicity and degradation of pollutant under both UV and visible light was observed.

Nanofiltration for Pharmaceutical Removal from Water

The release of wastewater treatment plant (WWTP) effluent can be considered a major source of pollution in drinking water sources, along with contamination from landfill leachate, sewage system or septic tank leakage, and contamination from agricultural runoff, as surface and groundwater account for a significant proportion of drinking water sources. One group of compounds that has attracted increasing interest in recent years is that of so called "contaminants of emerging concern," such as pharmaceuticals and personal care products (PPCPs) or hazardous organic pollutants (HOPs). Concentration of PPCPs in drinking water sources will be typically at least an order of magnitude less than that observed in WWTP effluents, due to the processes of dilution and sorption to solids, but degradation of these compounds is unlikely [65].

Many PPCPs have been implicated as having endocrine disrupting properties, affecting hormone levels and reproductive function of aquatic animals, promoting the increase in antibiotic resistance of microorganisms, and other both chronic and acute health effects. While levels found in drinking water sources are typically very low (of the ng/L to μg/L order of magnitude), the effects of long term exposure to low levels of these chemicals is unknown [66, 67]. In addition, certain compounds such as Diclofenac have been found in mg/L levels [68]. Many of the transformation products or metabolites of these substances can themselves be considered contaminants of concern [69]. A number of studies on the occurrence of PPCPs in various matrices have shown that a wide variety of pharmaceuticals at varying concentrations occur in WWTP effluents [70], WWTP influent and effluent streams [71, 72], in sewage sludge and in sludge enriched soils [73] and in the marine and wider aquatic matrices [74, 75]. In some cases (e.g., carbamazepine) the effluent concentrations from WWTPs was found to be higher than the influent concentration [71, 76]. WWTP effluents containing PPCPs are a good indication that the local drinking water sources may be contaminated. Many drinking water treatment plants do not have the technical capacity to remove micropollutants such as pharmaceuticals and other trace organic compounds [77]. Although adverse health effects from long term, low level exposure to these compounds is not yet proven, drinking water treatment facilities should be designed to provide water that is free from pharmaceuticals [78].

Regulation standards for drinking water (including that from reclaimed water) are designed to ensure that the water is of a reliable and safe quality. However, discharge guidelines from wastewater treatment plants (WWTPs) for many PPCPs are not yet encompassed in the common regulatory frameworks, likewise there remains a lack of regulation for limits in drinking water. The European Water Framework Directive (WFD) [Directives 2000/60/EC, 2008/105/EC and 2009/90/EC] contains a dynamic list of priority pollutants which should be regulated and monitored in WWTP effluents [70]. 12 new substances have recently been added to the original 33 priority pollutants and 8 other pollutants, with a new requirement for monitoring and reporting of 3 pharmaceutical compounds (an anti-inflammatory drug, diclofenac, and two hormones, 17α-ethinylestradiol and 17β-estradiol) [74]. These compounds now appear on the EPA Watch List, with the potential for future addition to the list of priority pollutants. In order to assure water quality and future proof the potential for reclaimed water use for potable purposes, the development and application of treatment technologies that are capable of removing PPCPs is critical.

Typical drinking water treatment plants consist of a number of stages, including chemical coagulation/settling, sand filtration, adsorption using granular activated carbon (GAC) and disinfection using chlorination, ozone, UV or a combination of these technologies. Chemical coagulation has shown limited potential for removal of pharmaceuticals, reducing the concentrations of ionisable pharmaceuticals only but not removing them completely, while non ionisable pharmaceuticals such as carbamazepine were unaffected [76, 79]. Sand filtration has been shown to be ineffective in the removal of pharmaceuticals such as bezafibrate, clofibric acid, carbamazepine and diclofenac, while high dose ozone (3 mg/L) was not capable of removing clofibric acid [80]. UV was shown to be ineffective in the removal of antibiotics, although chlorination and ozone were able to reduce concentrations to a reasonable level [81]. Residual free chlorine in the drinking water treatment system was shown to be capable of further degradation of approximately 50% of a cohort of 98 studied pharmaceuticals in treated drinking water [82]. However, due to the nature of many pharmaceuticals (polar, with acidic or basic functional groups), ozonation or chlorination can result in transformation of compounds rather than removal [83]. Thus there is a driver for physical removal rather than transformation, a function which membrane separation processes such as nanofiltration could reasonably perform. There is very little research on the application of graphene enhanced nanofiltration membranes on the removal of PPCPs from drinking or wastewater streams, and given the promising results in other areas such as desalination, this is an area that is worthy of increased attention. The background to the issue and the limited research in the area is presented in this section.

While reverse osmosis (RO) is an established unit operation in the advanced treatment of water, the application of nanofiltration (NF) is a membrane separation process that is gaining increased attention in the literature for the removal of trace organic contaminants from drinking water. Nanofiltration is a medium-to-high pressure (7 – 40 bar) crossflow membrane filtration process employing membranes with pore sizes in the range 0.5 – 5 nm and molecular weight cut-offs (MWCO) in the range 200 – 2000 Da. It has been shown to remove protozoa (including oocytes), bacteria and viruses, as well as natural organic matter (NOM), organohalides, pesticides and pharmaceuticals [84, 85]. NF compares favourably with RO, in that it can be operated at lower applied pressure and rejects fewer mineral ions – meaning that the energy requirements are less, and that the permeate needs less post-treatment [77, 86].

While it would appear at first glance that most pharmaceuticals, which have molecular weights within 150-500 Da, should be automatically rejected by NF membranes, NF does not operate simply on a size exclusion basis. Many characteristics of both the pharmaceuticals and the membranes themselves can affect solute rejection, including their polar/non-polar nature, acid disassociation constant, hydrophobicity/hydrophilicity, molecular weight and molecular size, and membrane charge. Typically the only information supplied by NF membrane manufacturers is the MWCO, and the molecular weight of a component which will have a 90% rejection, implying that larger molecules will have a rejection of more than 90%, while smaller molecules will have a rejection of less than 90% [87]. This, however, is often not the case, for example where size exclusion may dominate the rejection in one membrane type, rejection for another membrane material even with similar pore size may be dominated by electrostatic interactions [88], and there is a large variation in the literature between reported rejection/retention values for pharmaceuticals [89].

Rejection or retention of pharmaceuticals on NF membranes is governed by the operating parameters (e.g., crossflow velocity, operating temperature, transmembrane pressure), solute properties (e.g., charge, hydrophilic /hydrophobic, polar/non-polar), membrane properties (e.g., surface roughness, charge, pore size/MWCO, functional groups (acidic/basic)) and feed composition (e.g., pH, ionic strength, presence of other substances such as humic acids) [77, 90, 91]. The most well understood mechanism of rejection of solutes by NF is the physical sieving of solutes larger than the MWCO of the membrane, while steric effects will also lead to rejection. Other physico-chemical interactions such as adsorption and electrostatic diffusion are also important in rejection, as are charge exclusion effects (electrical and Donnan). Hydrophobic-hydrophobic interactions between the membrane and solute and diffusion limitation of the solute is also seen to play a role [41, 92, 93]. Steric hindrance is the most likely mechanism for rejection of uncharged, hydrophilic compounds, and correlation between the molecular size/weight and the membrane pore size distribution has been seen in the rejection of solutes of this type [87, 94].

Molecules that are cylindrical in shape can display a tendency to approach the membrane pores in a vertical manner, due to the polarity of the molecules (dipole moment) being attracted to fixed charge groups on the membrane surface. Thus low retention of high dipole moment organic compounds can be influenced by steric effects [95]. Molecules with low dipole moment such as diclofenac and ibuprofen have been shown to have greater rejection than molecules such as carbamazepine [96]. Less polar (low dipole moment) molecules can be preferentially adsorbed to the membrane, particularly if they are also hydrophobic. Adsorption typically results in a lower steady state rejection, and a time dependent effect in NF will be observed as the adsorption capacity of the membrane is reached [77]. Removal efficiency of solutes is also related to their octanol water partition coefficient ($\log K_{OW}$), which is related to their hydrophobicity, and interaction with hydrophobic membranes. High $\log K_{OW}$ (compounds such as diazepam, ibuprofen and diclofenac) and low solubility can lead to an increase in rejection, and this in turn can in part be attributed to hydrophobic adsorption to the membrane surface [97, 98]. Ibuprofen, for example, in its neutral form, has a relatively high hydrophobicity, and reasonably high levels of adsorption to hydrophobic membrane surfaces will typically be observed. In the presence of other solutes which also adsorb to the membrane surface (competitive adsorption), it may be observed that an increase in rejection occurs for compounds that sorb less in comparison with other compounds within a mixed matrix, in comparison to the rejection that would be observed in a single-component feed [99, 100]. In addition, adsorption can lead to a decrease in the time required to achieve steady state rejection, leading to overestimation of the short-term rejection capabilities of membranes even for compounds with a low $\log K_{OW}$ [101, 102].

Electrostatic repulsion is seen to play a role, where rejection of solutes with the same charge as the membrane is enhanced [41, 103]. For negatively charged membranes, negatively charged solutes are better removed compared to uncharged solutes, while positively charged solutes demonstrate poor removal. However membrane shielding effects can be observed for solutes with the opposite charge to the membrane with increasing solute concentrations, due to charge shielding effects on the membrane surface [104, 105]. A "charge concentration effect" may be observed, where positively charged and neutral molecules engage in hydrophobic interactions with a negatively charged membrane, while negatively charged molecules cannot approach the membrane surface (Figure 7). Thus increased concentration of positively charged solutes will be seen at the membrane surface in comparison with the bulk fluid [95, 106].

The pH and pK_a values of the solutes have a significant effect on the retention of PPCPs on charged membranes, particularly for ionizable pharmaceuticals such as sulfamethoxazole and ibuprofen, whereas it is less important for non-ionizable compounds such as carbamazepine. Retention on negatively charged membranes can be seen to increase as the compound becomes negatively charged above its pK_a value, for example the rejection of amoxicillin was enhanced by 85% with increase in pH above the pK_a [107]. Rejection of ibuprofen (hydrophobic and acidic) at pHs lower than the pK_a led to reduction in diffusion through the membrane due to partial adsorption on a negatively charged membrane surface, while electrostatic repulsion was dominant at higher pH [108]. High ionic strengths can minimise the contribution of electrostatic interactions on rejection, by suppression of the double layer or Debye screening length, although this effect is thought to be not as significant, due to the formation of a hydrated layer around the charged functional groups of the pharmaceuticals [109]. The speciation of pharmaceuticals (i.e., the different forms in which they occur, and the transition between those forms) can also lead to change in retention as a function of pH [95]. For example, Clonazepam is protonated at highly acidic conditions and it becomes non-protonated (neutral) when the pH is increased up to 6. However, under alkaline conditions, the compound changes to its enolic form, which has an enhanced water affinity due to the charge on the molecule; this in turn leads to reduction in retention [97].

The mechanism of transport through NF membranes is generally thought to be a combination of solution/diffusion processes, as well as flow through defects or pores in the nanometer scale on the membrane surface [99, 110, 111]. The concentration difference across the membrane is a driver for flux of solutes through the membrane by diffusion. It has been shown that an increase in water recovery, increasing the concentration differential across the membrane, leads to a decrease in rejection of solutes [112]. The solute and water transport through the membrane pores can be enhanced by increasing the cross-flow velocity or transmembrane pressure, and it has also been shown experimentally that this can lead to an increase in rejection due to a reduction in adsorption effects [113].

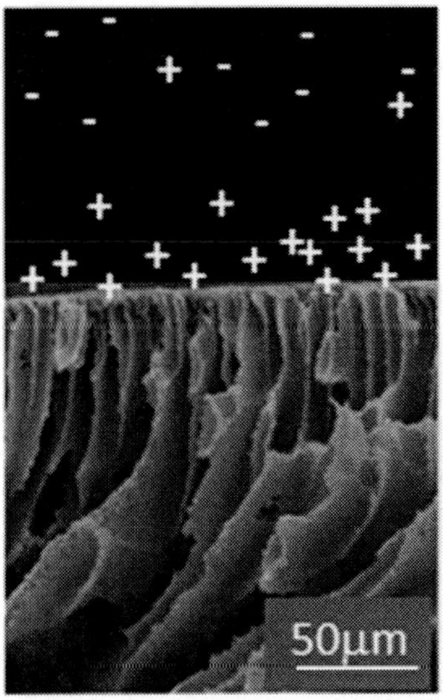

Figure 7. Charge concentration effect – SEM image of membrane with GO-TiO_2 (1 wt%)/PES porous support layer.

The ability to predict the performance of NF in rejection of PPCPs has been a subject of increasing research. Mathematical models have been developed that incorporate various combinations of feed, operating, membrane and solute parameters in an effort to develop robust predictions. Early attempts at modelling rejection in NF systems were based on either variations of Hermia models or Nernst-Planck equations for mass transfer. In modified Hermia models, fouling/rejection in larger pore size membranes was governed by both size exclusion and electrostatic effects while in smaller pore size membranes size or steric effects were assumed to dominate, and good agreement with prediction was found for the rejection of acetaminophen and sulfamethoxazole [114]. Models based on the extended Nernst-Planck equation for mass transfer were developed, taking the molecular size of solutes and hence hindered transport through narrow pores into account. This model accounted for convection, ionic diffusion and electromigration of species within the membrane pores. Donnan equilibrium was assumed to be the electrostatic interaction involved in ion partitioning between the membrane and liquid phases [115, 116]. Other research has focused on the development of membrane transport models, to elucidate the contributions of diffusion and convection on solute transport and rejection. Modelling studies have shown that diffusion is important for hydrophobic non-polar compounds while convection dominates for other molecules, and that convection is more important for loose NF membranes [92]. Analytical models for solute transport and rejection have also been developed, such as those incorporating the free energy of interaction and the size exclusion contribution, with good agreement between prediction and experiment achieved for a range of hydrophilic/hydrophobic/charged/uncharged pharmaceuticals achieved by Botton et al. [42].

This type of model was further developed by inclusion of the dielectric potential as part of the ion partitioning effect, termed the Donnan Steric Pore Model and Dielectric Effect (DSPM&DE) [117]. This model was deemed quite simple to use, involving just three adjustable parameters: the surface charge density, average pore size, and effective membrane thickness, and was applied in prediction of rejection of hormones and some PPCPs. The inclusion of dielectric effect enhanced the realistic behaviour predicted by the model [118]. However the rejection was often overestimated, due to the underestimation of the contribution of hydrophobic interactions with the membrane [95]. The model was refined to take this affinity effect into account, and predictions were improved for hydrophilic compounds including acyclovir, caffeine and ranitidine, however retention of hydrophobic molecules was still over-estimated [77], despite extensive model verification for salts [119]. This model was incorporated into a freely available computer programme, NanoFiltran, developed by Geraldes & Brites Alves [120] and further developed by incorporation of the increase in solute concentration along the membrane module for seawater desalination [121], however it has not been applied to PPCPs in the literature. While the DSPM&DE type models are suited for charged solutes, models based on the Spiegler-Kedem-Katchalsky equations (SKK) have been successfully applied to neutral molecules, including the pharmaceuticals isopropylantipyrine and antipyrine [122]. A quasi-empirical model combining predictions for neutral and charged solutes (taking pH dependency into account) was applied reasonably successfully for carpamazepine, ibuprofen, sufadiazine, sulfamethoxazone and sulfamethazone

[54]. This simplified charge concentration polarization (SCCP) model can predict the effects of electrostatic interactions on the rejection of trace organic solutes [106], however without the presence of charge interactions (e.g., at pH lower than pK_a), experimental measurements must be incorporated into the model.

The effect of organic fouling on membrane characteristics is typically found to be membrane dependent, with pore size deemed to be the dominant factor. For example, fouled tight NF membranes have been shown to become more hydrophilic and negatively charged, whereas loose NF membranes became more hydrophobic and less negatively charged. The membrane physical characteristics also play a role, with hydrophobic membranes with high surface roughness being subject to increased levels of adsorption of organics over smooth, hydrophilic membranes [114, 123, 124]. The major compounds in drinking water sources that have the tendency to foul NF membranes include humic acids and polysaccharides, and model foulants to mimic typical organic fractions and colloidal materials in treated secondary wastewater and surface water are often approximated in laboratory studies using a combination of humic acids, bovine serum albumin, alginate and colloidal silica [125, 126]. A fouling layer can build up on the surface of the membrane and within the pores, and this can modify the membrane properties such as hydrophobicity and surface charge, and can also lead to cake enhanced concentration polarization [127]. Membrane fouling and its effect on retention of PPCPs is complex, and both enhancement and decreases in rejections have been reported in the literature. Fouling can depend on the characteristics of the membrane and on the characteristics of the dissolved matter in the feed, and synergistic effects may be observed in mixed foulant systems. Membrane properties such as surface roughness, pore size, and hydrophobicity can have an impact on the level of fouling and thus the level of flux decline that would result from this [78]. In addition to fouling by feed components, the solutes themselves can have an impact on fouling, for example flux decline has been attributed to adsorption of compounds such as diclofenac within the pores of loose NF membranes [96]. Fouling of loose NF membranes, where the mechanism of retention may be governed by pore restriction or blocking during fouled conditions, may lead to an increase in retention, while tight NF membranes may experience a cake enhanced concentration polarization effect, due to modification of the surface charge. Foulant deposition is dependent on feed characteristics, for example with calcium carbonate and sulfate depositing in the early stages of filtration and organic foulants gradually forming a dense packing layer subsequently [128]. Increased hydraulic resistance due to a mixed cake layer structure can enhance flux decline, with adsorption of organics playing a key role, and synergistic fouling can be observed, with more than single layer fouling leading to hindrance of back transport of foulants from the fouling layer [129]. Dissolved organic matter with a medium to low MW can result in membrane fouling by pore blocking, and those with strong hydrophobic characteristics can form a fouling layer, altering the hydrophobicity and surface charge of the membrane [130]. Increased retention of hydrophobic trace organics due to formation of a fouling layer reducing hydrophobic interactions between the solutes and the membrane surface may be observed, with a decrease in diffusional transport across the membrane [131], however, with the decrease in electrostatic interaction based rejection, and change in the surface free energy between membrane and foulants, cake enhanced concentration polarization can lead to decreased rejection [130], where the higher concentration next to the membrane increases the mass transport by diffusion. This has been observed for hydrophilic compounds such as sulfamethoxazole, ibuprofen and carbamazepine with organic fouling on loose NF membranes. The presence of calcium ions leads to increased flux decline, however the cake layers formed can be quite cross-linked and porous, with low hydraulic resistance when alginate or humic acids were the basis of the fouling layers [125]. Calcium ions and humic acid fouling can positively influence the retention of hydrophilic and neutral compounds, where size exclusion is dominant, however concentration polarization decreases the rejection of larger molecules [114]. Biofouling, where microbial layers can build up on surfaces, is another type of membrane fouling that can impact on flux and rejection of PPCPs [42].

Membrane scaling can often occur, particularly in water recycling applications. Scale forming inorganic salts such as $CaSO_4$ and $CaCO_3$ can interact with organic matter, and have an effect on the flux and rejection characteristics. Humic acids and other natural organic matter can act as anti-scalants, and may act to prevent the expected decline in flux, by forming a porous cake layer. The membrane scaling onset may be delayed by organic matter, leading to a delay in the decline in rejection of PPCPs; however the decline in rejection once scaling begins can be very severe, due to the combined cake-enhanced concentration polarization as a result of the combined membrane fouling and scaling, with the scaling layer leading to an additional concentration polarization effect and reducing shear effects within the underlying fouling layer [89, 132–134].

Improved rejection of pharmaceuticals can be achieved by tailored fabrication of membranes or modification of commercially available NF membranes. Enhanced rejection by modification of support layer by polymer blending has been investigated [107], while the incorporation of molecules to enhance charge or hydrophilicity is a common approach for improvement of separation capabilities and fouling reduction [135]. The use of charged molecules within the membrane matrix will generally lead to superior removal of charged molecules such as ibuprofen and sulfamethoxazole but poor removal of molecules such as carbamazepine [136]. Graphene-containing ceramic composite tubular membranes were prepared by dip-coating of a graphene-TiO_2 mixture onto tubular Al_2O_3 membranes, followed by calcination. The removal of three pharmaceuticals (cephalexin (CLX), sulfamethoxazole (SMX) and caffeine (CAF)) was investigated under simultaneous electrocoagulation/electrofiltration (EC/EF). CLX and SMX have pKa values of 4.5 and 6.2, and zeta potentials of – 6.14 mV and – 14.45 mV respectively; this resulted in their negative charge in aqueous solution leading to rejection by electrostatic repulsion from the negatively charged graphene containing membrane, with rejections of 23 and 46% respectively under standard crossflow conditions in model water, with significantly higher removal achieved in real WWTP influent; significant improvement in removal from model water was seen using EC/EF technique. Removal from WWTP influent was almost complete without EC/EF in any case due to adsorption of pharmaceuticals onto NOM, however permeate flux was improved with increase in applied electric field strength. The pK_a of CAF is 10.4 with zeta potential of – 0.77 mV, and as such very poor removal was seen.

Integrated systems for PPCP removal have been explored extensively in the literature, including the use of NF with advanced oxidation processes (AOPs) such as ozone, peroxide, UV, photo-Fenton, photocatalysis, electrocoagulation [137] and electrochemical AOPs [138], and have in many cases shown enhanced performance of NF in terms of enhanced flux and removal of organic contaminants [139]. Ozonation for example can be efficient in the removal of polar compounds such as diclofenac and carbamazepine, although not effective on clofibric acid or bezafibrate [80]. While granular activated carbon (GAC) operations alone are not particularly useful for removal of hydrophilic, polar/charged high MW compounds (including clofibric acid, diclofenac and carbamazepine), the inclusion of an NF step prior to GAC removes the majority of NOM that is present for competitive adsorption, leading to enhanced removal of PPCPs by GAC [104]. The adsorption capacity of GO and rGO has also shown promise in the removal of PPCPs, with membranes prepared for adsorptive removal of cations, anions and BPA developed with selective adsorption characteristics on UF membranes [140]. A photocatalytic cubic Ag/AgBr/GO nanocomposite showed complete degradation of diclofenac within 6 minutes under visible light irradiation [141] – integration of this type of composite within a membrane would be of great interest.

Nanofiltration can be suitable for removal of virtually all pharmaceuticals and trace organic compounds. However, a single nanofiltration membrane will not provide a complete solution to all potential PPCP contamination. For example, removal of a wide range of PPCPs from real surface waters was found to range from 0% to over 90% [90], with poor rejection often being associated with looser NF membranes and uncharged low MW molecules [77], while studies testing novel membranes incorporating surface charge modifying molecules, unsurprisingly, show good removal of molecules like ibuprofen but poor removal of molecules such as carbamazepine [136]. However, the removal of PPCPs by NF is generally above that which is possible with conventional treatment [142]. The solute characteristics and the membrane characteristics must be carefully matched, and operating parameters such as pH, ionic strength and crossflow velocity must be appropriate. It should be possible to develop an integrated system, for example in which NF membranes with different characteristics are applied in series, to provide a complete solution to PPCP contamination in drinking water treatment. However there are two issues with this – the energy costs associated with extensive NF, and the subsequent treatment of NF concentrate. The components of NF concentrate from drinking water treatment have typically been subjected previously to a range of operations from WWTP processes, or adsorption/advection during migration to the drinking water source. As such they are generally difficult to treat or biodegrade, and appropriate treatment processes for the concentrate must also be developed, such as UV/ozone treatment [143]. However, although subsequent treatment of concentrate is required, there can be savings in overall materials or energy costs, by application of advanced processes such as solar photo-Fenton or UV/O_3 to the smaller volume concentrate rather than to the full volumetric flow of water [143, 144], and overall an integrated system is recommended [145].

Proton Exchange Membranes for Fuel Cells (PEMFCs)

Due to increasing demand for alternative energy sources, research and development continues to increase into fuel cells. The two electrodes in a fuel cell are separated by a membrane, which acts as the electrolyte and allows the passage of ions between the electrodes (Figure 8). Ion exchange membranes are a key area for innovation, as the high cost of membranes such as Nafion® and others remains a barrier to large scale implementation [146], and so-called Proton Conducting – Mixed Matrix Membranes (PC-MMM) have the potential to meet this challenge. In particular, mixed matrix proton exchange membrane fuel cells show great promise due to their small size, relatively low cost of manufacture, and high efficiency.

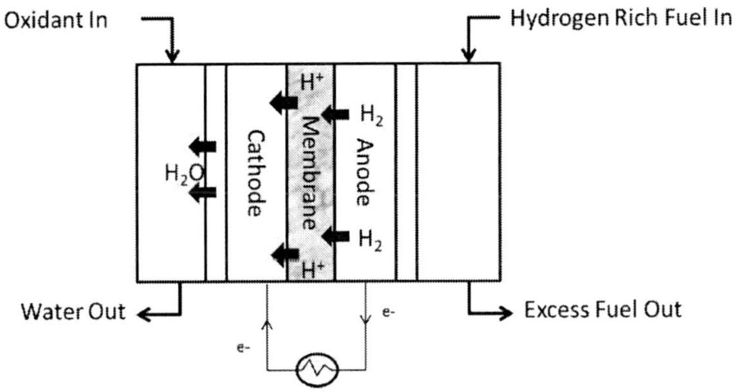

Figure 8. PEMFC Schematic.

Although GO is an electronic insulator, multilayer GO in particular exhibits superfast conduction of protons due to their ability to utilise the hydrogen bonding networks along the adsorbed water layers that are formed on GO surfaces, with epoxide groups being particularly important for proton transport [147, 148]. Ozonation of GO sheets has also been shown to increase proton transport, due to higher oxygen containing functional groups and morphology changes in the GO sheets [149].

Carbon nanosheets such as GO have been incorporated into polymeric membranes for proton exchange, with associated improvements to the thermal and mechanical stability and electrochemical properties. In particular, use of membrane materials such as Nafion® have been shown to be problematic in terms of diffusion of fuel (e.g., methanol) in addition to protons across the membrane [150]. Extensive tuning of GO nanosheet properties is possible to address this, with increase in GO flake size in lamellar GO membranes leading to a corresponding decrease in methanol permeation, without any impact on the proton conductivity [151]. Membranes can be fabricated using a sol-gel process, and by the inclusion of carbon nanosheets, the structure and performance of the membrane can be enhanced. The inclusion of GO in this manner was shown to substantially decrease the cross-membrane diffusion of methanol molecules in direct methanol fuel cells (DMFCs) while still allowing good proton transport [146]. Zwitterion-coated graphene oxide was incorporated into polybenzimidazole solution cast membranes, yielding low methanol permeability and high proton conductivity comparable to Nafion 117 [150], although temperature stability was slightly reduced. GO has also been immobilized on Nafion® by atom transfer radical addition, leading to a nanohybrid membrane with a 35% increase in fuel cell performance and a 1.6 fold improvement in proton conductivity [152].

Conclusion

The incorporation of graphene based nanosheets into membranes shows great promise in enhancement of flux (for example water flux in desalination or water purification) or proton flux in fuel cells. The extremely high capacity for water transport, high hydrophilicity and tunable gap spacing of GO sheets coupled with their ease and low cost of manufacture indicates that their incorporation into a widening variety of membrane type will continue to increase. Graphene based nanosheets have been incorporated into polymeric membranes for liquid phase separation by mixing of GO/rGO with polymer followed by membrane fabrication using standard methods such as phase separation or thin film composite preparation. Graphene/GO has also been immobilized onto ceramic membranes by vacuum filtration, spin casting or drop coating, and incorporated into electrospun polymeric membranes. The enhanced hydrophilicity and decrease in surface roughness typically improves the flux and can enhance the separation efficiency due to increase in membrane charge for electrostatic exclusion. GO/rGO can enhance photocatalytic activity when incorporated into nanocomposite membranes with photocatalyic molecules such as TiO_2, and nanofiller molecules can also act to tune the spacing between nanosheets allowing for manipulation of selectivity. GO has been shown to have excellent proton conducting capacity and is gaining increasing interest as a membrane modification material for incorporation into fuel cells, in particular direct methanol fuel cells, where GO functionalization of proton exchange membranes can help to solve the problem of undesirable methanol transport.

Graphene/GO/rGO functionalized membranes have exhibited superiority over conventional membrane processes in a wide range of applications, and the number of research papers into graphene functionalized membranes has increased exponentially in the last few years. The potential for commercialization and wide scale use of these types of membranes has massive potential with many applications yet to be discovered.

ACKNOWLEDGMENTS

This work was supported by funding from the Environmental Protection Agency (EPA Ireland) under project 2011-W-MS-8. Thanks are due to Mr. Declan McGlade, PhD student funded under this project, for some of the images contained in this chapter.

REFERENCES

[1] Nair, R. R.; Wu, H. A.; Jayaram, P. N.; Grigorieva, I. V; Geim, A. K. Unimpeded permeation of water through helium-leak-tight graphene-based membranes. *Science* 2012, *335* (6067), 442–444.

[2] Sun, P.; Wang, K.; Zhu, H. Recent Developments in Graphene-Based Membranes: Structure, Mass-Transport Mechanism and Potential Applications. *Adv. Mater.* 2016, *28* (12), 2287–2310.

[3] Dreyer, D. R.; Park, S.; Bielawski, C. W.; Ruoff, R. S. The chemistry of graphene oxide. *Chem. Soc. Rev.* 2010, *39* (1), 228–240.

[4] Mi, B. Graphene oxide membranes for ionic and molecular sieving. *Science* 2014, *343* (6172), 740–742.

[5] Yoshida, H.; Bocquet, L. Labyrinthine water flow across multilayer graphene-based membranes: Molecular dynamics versus continuum predictions. *J. Chem. Phys.* 2016, *144* (23), 234701.

[6] Yin, J.; Deng, B. Polymer-matrix nanocomposite membranes for water treatment. *J. Memb. Sci.* 2015, *479*, 256–275.

[7] Kumar, M.; McGlade, D.; Ulbricht, M.; Lawler, J. Quaternized polysulfone and graphene oxide nanosheet derived low fouling novel positively charged hybrid ultrafiltration membranes for protein separation. *RSC Adv.* 2015, *5* (63), 51208–51219.

[8] Baker, R. W. *Membrane Technology and Applications*, 2nd ed.; Wiley: West Sussex, UK, 2004.

[9] Cheryan, M. *Ultrafiltration and Microfiltration Handbook*, 2nd ed.; CRC Press: Florida, USA, 1998.

[10] Schäfer, A. I.; Fane, A. G.; Waite, T. D. *Nanofiltration*; Elsevier, 2006.

[11] Foley, G. *Membrane Filtration: a Problem Solving Approach with MATLAB®*, 1st ed.; Cambridge University Press: Dublin, Ireland, 2013.

[12] Ghorbani, M.; Abdizadeh, H.; Golobostanfard, M. R. Reduction of Graphene Oxide via Modified Hydrothermal Method. *Procedia Mater. Sci.* 2015, *11*, 326–330.

[13] Krishnamoorthy, K.; Veerapandian, M.; Yun, K.; Kim, S.-J. The chemical and structural analysis of graphene oxide with different degrees of oxidation. *Carbon N. Y.* 2013, *53*, 38–49.

[14] Hu, M.; Mi, B. Enabling Graphene Oxide Nanosheets as Water Separation Membranes. 2013.

[15] Xu, C.; Cui, A.; Xu, Y.; Fu, X. Graphene oxide–TiO2 composite filtration membranes and their potential application for water purification. *Carbon N. Y.* 2013, *62*, 465–471.

[16] Kumar, M.; Gholamvand, Z.; Morrissey, A.; Nolan, K.; Ulbricht, M.; Lawler, J. Preparation and characterization of low fouling novel hybrid ultrafiltration membranes based on the blends of GO−TiO2 nanocomposite and polysulfone for humic acid removal. *J. Memb. Sci.* 2016, *506*, 38–49.

[17] Ganesh, B. M.; Isloor, A. M.; Ismail, A. F. Enhanced hydrophilicity and salt rejection study of graphene oxide-polysulfone mixed matrix membrane. *Desalination* 2013, *313*, 199–207.

[18] Kang, S.; Herzberg, M.; Rodrigues, D. F.; Elimelech, M. Antibacterial effects of carbon nanotubes: size does matter! *Langmuir* 2008, *24* (13), 6409–6413.

[19] Feng, B.; Xu, K.; Huang, A. Covalent synthesis of three-dimensional graphene oxide framework (GOF) membrane for seawater desalination. *Desalination* 2016, *394*, 123–130.

[20] Nazarpour, S.; Waite, S. R. *Graphene Technology: From Laboratory to Fabrication*, 1st ed.; Wiley, 2016.

[21] Sun, P.; Zhu, M.; Wang, K.; Zhong, M.; Wei, J.; Wu, D.; Xu, Z.; Zhu, H. Selective Ion Penetration of Graphene Oxide Membranes. *ACS Nano* 2013, *7* (1), 428–437.

[22] Huang, H.; Song, Z.; Wei, N.; Shi, L.; Mao, Y.; Ying, Y.; Sun, L.; Xu, Z.; Peng, X. Ultrafast viscous water flow through nanostrand-channelled graphene oxide membranes. *Nat. Commun.* 2013, *4*.

[23] Safarpour, M.; Vatanpour, V.; Khataee, A.; Esmaeili, M. Development of a novel high flux and fouling-resistant thin film composite nanofiltration membrane by embedding reduced graphene oxide/TiO2. *Sep. Purif. Technol.* 2015, *154*, 96–107.

[24] Hegab, H. M.; Zou, L. Graphene oxide-assisted membranes: Fabrication and potential applications in desalination and water purification. *J. Memb. Sci.* 2015, *484*, 95–106.

[25] Surwade, S. P.; Smirnov, S. N.; Vlassiouk, I. V.; Unocic, R. R.; Veith, G. M.; Dai, S.; Mahurin, S. M. Water desalination using nanoporous single-layer graphene. *Nat. Nanotechnol.* 2015, *10* (5), 459–464.

[26] Celebi, K.; Buchheim, J.; Wyss, R. M.; Droudian, A.; Gasser, P.; Shorubalko, I.; Kye, J.-I.; Lee, C.; Park, H. G. Ultimate permeation across atomically thin porous graphene. *Science* 2014, *344* (6181), 289–292.

[27] Abdelhalim, A. O. E.; Galal, A.; Hussein, M. Z.; El Sayed, I. E.-T. Graphene Functionalization by 1,6-Diaminohexane and Silver Nanoparticles for Water Disinfection. *J. Nanomater.* 2016, *2016*, 1–7.

[28] Roy, I.; Rana, D.; Sarkar, G.; Bhattacharyya, A.; Saha, N. R.; Mondal, S.; Pattanayak, S.; Chattopadhyay, S. Physical and electrochemical characterization of reduced graphene oxide/silver nanocomposites synthesized by adopting a green approach. *RSC Adv.* 2015, *5* (32), 25357–25364.

[29] Liu, Q.; Xu, G.-R. Graphene oxide (GO) as functional material in tailoring polyamide thin film composite (PA-TFC) reverse osmosis (RO) membranes. *Desalination* 2016, *394*, 162–175.

[30] Manawi, Y.; Kochkodan, V.; Hussein, M. A.; Khaleel, M. A.; Khraisheh, M. Can carbon-based nanomaterials revolutionize membrane fabrication for water treatment and desalination? *Desalination* 2016, *391*, 69–88.

[31] Mansourpanah, Y.; Shahebrahimi, H.; Kolvari, E. PEG-modified GO nanosheets, a desired additive to increase the rejection and antifouling characteristics of polyamide thin layer membranes. *Chem. Eng. Res. Des.* 2015, *104*, 530–540.

[32] Shao, L.; Chang, X.; Zhang, Y.; Huang, Y.; Yao, Y.; Guo, Z. Graphene oxide cross-linked chitosan nanocomposite membrane. *Applied Surface Science*. 2013, pp 989–992.

[33] Choi, W.; Choi, J.; Bang, J.; Lee, J. H. Layer-by-layer assembly of graphene oxide nanosheets on polyamide membranes for durable reverse-osmosis applications. 2013, *5* (23), 12510–12519.

[34] Hu, M.; Mi, B. Layer-by-layer assembly of graphene oxide membranes via electrostatic interaction. *J. Memb. Sci.* 2014, *469*, 80–87.

[35] Sun, P.; Chen, Q.; Li, X.; Liu, H.; Wang, K.; Zhong, M.; Wei, J.; Wu, D.; Ma, R.; Sasaki, T.; Zhu, H. Highly efficient quasi-static water desalination using monolayer graphene oxide/titania hybrid laminates. *NPG Asia Mater.* 2015, *7* (2), e162.

[36] Park, M. J.; Phuntsho, S.; He, T.; Nisola, G. M.; Tijing, L. D.; Li, X.-M.; Chen, G.; Chung, W.-J.; Shon, H. K. Graphene oxide incorporated polysulfone substrate for the fabrication of flat-sheet thin-film composite forward osmosis membranes. *J. Memb. Sci.* 2015, *493*, 496–507.

[37] Zhao, H.; Chen, S.; Quan, X.; Yu, H.; Zhao, H. Integration of microfiltration and visible-light-driven photocatalysis on g-C3N4 nanosheet/reduced graphene oxide membrane for enhanced water treatment. *Appl. Catal. B Environ.* 2016, *194*, 134–140.

[38] Agboola, O.; Maree, J.; Mbaya, R. Characterization and performance of nanofiltration membranes. *Environ. Chem. Lett.* 2014, *12* (2), 241–255.

[39] Misdan, N.; Lau, W. J.; Ong, C. S.; Ismail, A. F.; Matsuura, T. Study on the thin film composite poly(piperazine-amide) nanofiltration membranes made of different polymeric substrates: Effect of operating conditions. *Korean J. Chem. Eng.* 2015, *32* (4), 753–760.

[40] Huang, H.; Mao, Y.; Ying, Y.; Liu, Y.; Sun, L.; Peng, X. Salt concentration, pH and pressure controlled separation of small molecules through lamellar graphene oxide membranes. *Chem. Commun.* 2013, *49* (53), 5963.

[41] Bellona, C.; Drewes, J. E.; Xu, P.; Amy, G. Factors affecting the rejection of organic solutes during NF/RO treatment--a literature review. *Water Res.* 2004, *38* (12), 2795–2809.

[42] Botton, S.; Verliefde, A. R. D.; Quach, N. T.; Cornelissen, E. R. Surface characterisation of biofouled NF membranes: role of surface energy for improved rejection predictions. *Water Sci. Technol.* 2012, *66* (10), 2122–2130.

[43] Nicolaï, A.; Sumpter, B. G.; Meunier, V. Tunable water desalination across graphene oxide framework membranes. *Phys. Chem. Chem. Phys.* 2014, *16* (18), 8646-8654.

[44] Cohen-Tanugi, D.; Grossman, J. C. Water Desalination across Nanoporous Graphene. *Nano Lett.* 2012, *12* (7), 3602-3608

[45] Thomas, M.; Corry, B.; Hilder, T. A. What Have We Learnt About the Mechanisms of Rapid Water Transport, Ion Rejection and Selectivity in Nanopores from Molecular Simulation? *Small* 2014, *10* (8), 1453–1465.

[46] Han, Y.; Xu, Z.; Gao, C. Ultrathin Graphene Nanofiltration Membrane for Water Purification. *Adv. Funct. Mater.* 2013, *23* (29), 3693–3700.

[47] Han, Y.; Jiang, Y.; Gao, C. High-Flux Graphene Oxide Nanofiltration Membrane Intercalated by Carbon Nanotubes. *ACS Appl. Mater. Interfaces* 2015, *7* (15), 8147–8155.

[48] Chen, X.; Qiu, M.; Ding, H.; Fu, K.; Fan, Y. A reduced graphene oxide nanofiltration membrane intercalated by well-dispersed carbon nanotubes for drinking water purification. *Nanoscale* 2016, *8* (10), 5696–5705.

[49] Chong, J. Y.; Aba, N. F. D.; Wang, B.; Mattevi, C.; Li, K. UV-Enhanced Sacrificial Layer Stabilised Graphene Oxide Hollow Fibre Membranes for Nanofiltration. *Sci. Rep.* 2015, *5*, 15799.

[50] Goh, K.; Setiawan, L.; Wei, L.; Si, R.; Fane, A. G.; Wang, R.; Chen, Y. Graphene oxide as effective selective barriers on a hollow fiber membrane for water treatment process. *J. Memb. Sci.* 2015, *474*, 244–253.

[51] Bano, S.; Mahmood, A.; Kim, S.-J.; Lee, K.-H. Graphene oxide modified polyamide nanofiltration membrane with improved flux and antifouling properties. *J. Mater. Chem. A* 2015, *3* (5), 2065–2071.

[52] Wang, J.; Gao, X.; Wang, J.; Wei, Y.; Li, Z.; Gao, C. O-(Carboxymethyl)-chitosan Nanofiltration Membrane Surface Functionalized with Graphene Oxide Nanosheets for Enhanced Desalting Properties. *ACS Appl. Mater. Interfaces* 2015, *7* (7), 4381–4389.

[53] Zhu, J.; Tian, M.; Hou, J.; Wang, J.; Lin, J.; Zhang, Y.; Liu, J.; Van der Bruggen, B. Surface zwitterionic functionalized graphene oxide for a novel loose nanofiltration membrane. *J. Mater. Chem. A* 2016, *4* (5), 1980–1990.

[54] Lin, Y.-L.; Lee, C.-H. Elucidating the Rejection Mechanisms of PPCPs by Nanofiltration and Reverse Osmosis Membranes. *Ind. Eng. Chem. Res.* 2014, *53* (16), 6798–6806.

[55] Wang, Q.; Lu, Y.; Li, N. Preparation, characterization and performance of sulfonated poly(styrene-ethylene/butylene-styrene) block copolymer membranes for water desalination by pervaporation. *Desalination* 2016, *390*, 33–46.

[56] Xu, K.; Feng, B.; Zhou, C.; Huang, A. Synthesis of highly stable graphene oxide membranes on polydopamine functionalized supports for seawater desalination. *Chem. Eng. Sci.* 2016, *146*, 159–165.

[57] Bhadra, M.; Roy, S.; Mitra, S. Desalination across a graphene oxide membrane via direct contact membrane distillation. *Desalination* 2016, *378*, 37–43.

[58] Pastrana-Martínez, L. M.; Morales-Torres, S.; Kontos, A. G.; Moustakas, N. G.; Faria, J. L.; Doña-Rodríguez, J. M.; Falaras, P.; Silva, A. M. T. TiO_2, surface modified TiO_2 and graphene oxide-TiO_2 photocatalysts for degradation of water pollutants under near-UV/Vis and visible light. *Chem. Eng. J.* 2013, *224*, 17–23.

[59] Rao, G.; Zhang, Q.; Zhao, H.; Chen, J.; Li, Y. Novel titanium dioxide/iron (III) oxide/graphene oxide photocatalytic membrane for enhanced humic acid removal from water. *Chem. Eng. J.* 2016, *302*, 633–640.

[60] Gao, Y.; Hu, M.; Mi, B. Membrane surface modification with TiO_2–graphene oxide for enhanced photocatalytic performance. *J. Memb. Sci.* 2014, *455*, 349–356.

[61] Gao, P.; Liu, Z.; Tai, M.; Sun, D. D.; Ng, W. Multifunctional graphene oxide–TiO_2 microsphere hierarchical membrane for clean water production. *Appl. Catal. B Environ.* 2013, *138*, 17–25.

[62] Pastrana-Martínez, L. M.; Morales-Torres, S.; Papageorgiou, S. K.; Katsaros, F. K.; Romanos, G. E.; Figueiredo, J. L.; Faria, J. L.; Falaras, P.; Silva, A. M. T. Photocatalytic behaviour of nanocarbon–TiO_2 composites and immobilization into hollow fibres. *Appl. Catal. B Environ.* 2013, *142*, 101–111.

[63] Almeida, N. A.; Martins, P. M.; Teixeira, S.; da Silva, J. A.; Sencadas, V.; Kühn, K.; Cuniberti, G.; Lanceros-Mendez, S.; Marques, P. A. A. P. TiO2/graphene oxide immobilized in P(VDF-TrFE) electrospun membranes with enhanced visible-light-induced photocatalytic performance. *J. Mater. Sci.* 2016, *51* (14), 6974–6986.

[64] Athanasekou, C. P.; Morales-Torres, S.; Likodimos, V.; Romanos, G. E.; Pastrana-Martinez, L. M.; Falaras, P.; Dionysiou, D. D.; Faria, J. L.; Figueiredo, J. L.; Silva, A. M. T. Prototype composite membranes of partially reduced graphene oxide/TiO_2 for photocatalytic ultrafiltration water treatment under visible light. *Appl. Catal. B Environ.* 2014, *158*, 361–372.

[65] Gómez, M. J.; Herrera, S.; Solé, D.; García-Calvo, E.; Fernández-Alba, A. R. Spatio-temporal evaluation of organic contaminants and their transformation products along a river basin affected by urban, agricultural and industrial pollution. *Sci. Total Environ.* 2012, *420*, 134–145.

[66] Nicolopoulou-Stamati, P.; Hens, L.; Sasco, A. J. Cosmetics as endocrine disruptors: are they a health risk? *Rev. Endocr. Metab. Disord.* 2015, *16*, 373–383.

[67] Sweeney, M. F.; Hasan, N.; Soto, A. M.; Sonnenschein, C. Environmental endocrine disruptors: Effects on the human male reproductive system. *Rev. Endocr. Metab. Disord.* 2015, *16*, 341–357.

[68] Praskova, E.; Plhalova, L.; Chromcova, L.; Stepanova, S.; Bedanova, I.; Blahova, J.; Hostovsky, M.; Skoric, M.; Maršálek, P.; Voslarova, E.; Svobodova, Z. Effects of subchronic exposure of diclofenac on growth, histopathological changes, and oxidative stress in zebrafish (Danio rerio). *ScientificWorldJournal.* 2014, *2014*, 645737.

[69] Evgenidou, E. N.; Konstantinou, I. K.; Lambropoulou, D. A. Occurrence and removal of transformation products of PPCPs and illicit drugs in wastewaters: A review. *Sci. Total Environ.* 2015, *505*, 905–926.

[70] Luo, Y.; Guo, W.; Ngo, H. H.; Nghiem, L. D.; Hai, F. I.; Zhang, J.; Liang, S.; Wang, X. C. A review on the occurrence of micropollutants in the aquatic environment and their fate and removal during wastewater treatment. *Sci. Total Environ.* 2014, *473-474*, 619–641.

[71] Lacey, C.; Basha, S.; Morrissey, A.; Tobin, J. M. Occurrence of pharmaceutical compounds in wastewater process streams in Dublin, Ireland. *Environ. Monit. Assess.* 2012, *184* (2), 1049–1062.

[72] Lacey, C.; McMahon, G.; Bones, J.; Barron, L.; Morrissey, A.; Tobin, J. M. An LC-MS method for the determination of pharmaceutical compounds in wastewater treatment plant influent and effluent samples. *Talanta* 2008, *75* (4), 1089–1097.

[73] Barron, L.; Purcell, M.; Havel, J.; Thomas, K.; Tobin, J.; Paull, B. Occurrence and Fate of Pharmaceuticals and Personal Care Products within Sewage Sludge and Sludge-Enriched Soils https://www.epa.ie/pubs/reports/research/waste/STRIVE_34_Barron_PCPs_web.pdf.pdf (accessed Feb 25, 2016).

[74] Quinn, B.; McEneff, G.; Schmidt, W. Pharmaceuticals in the Irish aquatic environment: the assessment and potential human impact of exposure to pharmaceuticals on marine and freshwater bivalves http://www.epa.ie/pubs/reports/research/water/Research Report 143 web.pdf (accessed Feb 25, 2016).

[75] McEneff, G.; Schmidt, W.; Quinn, B. Pharmaceuticals in the aquatic environment: A short summary of current knowledge and the potential impacts on aquatic biota and humans http://www.epa.ie/pubs/reports/research/water/research142pharmaceuticalsintheaquaticenvironment.html (accessed Feb 25, 2016).

[76] Vieno, N.; Tuhkanen, T.; Kronberg, L. Elimination of pharmaceuticals in sewage treatment plants in Finland. *Water Res.* 2007, *41* (5), 1001–1012.

[77] Wang, X.; Li, B.; Zhang, T.; Li, X. Performance of nanofiltration membrane in rejecting trace organic compounds: Experiment and model prediction. *Desalination* 2015, *370*, 7–16.

[78] Azaïs, A.; Mendret, J.; Gassara, S.; Petit, E.; Deratani, A.; Brosillon, S. Nanofiltration for wastewater reuse: Counteractive effects of fouling and matrice on the rejection of pharmaceutical active compounds. *Sep. Purif. Technol.* 2014, *133*, 313–327.

[79] Vieno, N.; Tuhkanen, T.; Kronberg, L. Removal of pharmaceuticals in drinking water treatment: effect of chemical coagulation. *Environ. Technol.* 2006, *27* (2), 183–192.

[80] Ternes, T. A.; Meisenheimer, M.; McDowell, D.; Sacher, F.; Brauch, H.-J.; Haist-Gulde, B.; Preuss, G.; Wilme, U.; Zulei-Seibert, N. Removal of Pharmaceuticals during Drinking Water Treatment. *Environ. Sci. Technol.* 2002, *36* (17), 3855–3863.

[81] Adams, C.; Wang, Y.; Loftin, K.; Meyer, M. Removal of Antibiotics from Surface and Distilled Water in Conventional Water Treatment Processes. *J. Environ. Eng.* 2002, *128* (3), 253–260.

[82] Gibs, J.; Stackelberg, P. E.; Furlong, E. T.; Meyer, M.; Zaugg, S. D.; Lippincott, R. L. Persistence of pharmaceuticals and other organic compounds in chlorinated drinking water as a function of time. *Sci. Total Environ.* 2007, *373* (1), 240–249.

[83] Snyder, S. A.; Westerhoff, P.; Yoon, Y.; Sedlak, D. L. Pharmaceuticals, Personal Care Products, and Endocrine Disruptors in Water: Implications for the Water Industry. *Environ. Eng. Sci.* 2003, *20* (5), 449–469.

[84] Shannon, M. A.; Bohn, P. W.; Elimelech, M.; Georgiadis, J. G.; Mariñas, B. J.; Mayes, A. M. Science and technology for water purification in the coming decades. *Nature* 2008, *452* (7185), 301–310.

[85] Van der Bruggen, B.; Vandecasteele, C. Removal of pollutants from surface water and groundwater by nanofiltration: overview of possible applications in the drinking water industry. *Environ. Pollut.* 2003, *122* (3), 435–445.

[86] Plumlee, M. H.; Stanford, B. D.; Debroux, J.-F.; Hopkins, D. C.; Snyder, S. A. Costs of Advanced Treatment in Water Reclamation. *Ozone Sci. Eng.* 2014, *36* (5), 485–495.

[87] Van der Bruggen, B.; Vandecasteele, C. Modelling of the retention of uncharged molecules with nanofiltration. *Water Res.* 2002, *36* (5), 1360–1368.

[88] Siegrist, H.; Joss, H. Review on the fate of organic micropollutants in wastewater treatment and water reuse with membranes. *Water Sci. Technol.* 2012, *66* (6), 1369–1376.

[89] Vogel, D.; Simon, A.; Alturki, A. A.; Bilitewski, B.; Price, W. E.; Nghiem, L. D. Effects of fouling and scaling on the retention of trace organic contaminants by a nanofiltration membrane: The role of cake-enhanced concentration polarisation. *Sep. Purif. Technol.* 2010, *73* (2), 256–263.

[90] Radjenović, J.; Petrović, M.; Ventura, F.; Barceló, D. Rejection of pharmaceuticals in nanofiltration and reverse osmosis membrane drinking water treatment. *Water Res.* 2008, *42* (14), 3601–3610.

[91] Simon, A.; Price, W. E.; Nghiem, L. D. Changes in surface properties and separation efficiency of a nanofiltration membrane after repeated fouling and chemical cleaning cycles. *Sep. Purif. Technol.* 2013, *113*, 42–50.

[92] Kim, T.-U.; Drewes, J. E.; Scott Summers, R.; Amy, G. L. Solute transport model for trace organic neutral and charged compounds through nanofiltration and reverse osmosis membranes. *Water Res.* 2007, *41* (17), 3977–3988.

[93] Darvishmanesh, S.; Buekenhoudt, A.; Degrève, J.; Van der Bruggen, B. Coupled series–parallel resistance model for transport of solvent through inorganic nanofiltration membranes. *Sep. Purif. Technol.* 2009, *70* (1), 46–52.

[94] Kosutic, K. Porosity of some commercial reverse osmosis and nanofiltration polyamide thin-film composite membranes. *J. Memb. Sci.* 2000, *168* (1-2), 101–108.

[95] Nghiem, L. D.; Schäfer, A. I.; Elimelech, M. Pharmaceutical Retention Mechanisms by Nanofiltration Membranes. *Environ. Sci. Technol.* 2005, *39* (19), 7698–7705.

[96] Vergili, I. Application of nanofiltration for the removal of carbamazepine, diclofenac and ibuprofen from drinking water sources. *J. Environ. Manage.* 2013, *127*, 177–187.

[97] Vona, A.; di Martino, F.; Garcia-Ivars, J.; Picó, Y.; Mendoza-Roca, J.-A.; Iborra-Clar, M.-I. Comparison of different removal techniques for selected pharmaceuticals. *J. Water Process Eng.* 2015, *5*, 48–57.

[98] Yoon, Y.; Westerhoff, P.; Snyder, S. A.; Wert, E. C. Nanofiltration and ultrafiltration of endocrine disrupting compounds, pharmaceuticals and personal care products. *J. Memb. Sci.* 2006, *270* (1-2), 88–100.

[99] Steinle-Darling, E.; Litwiller, E.; Reinhard, M. Effects of sorption on the rejection of trace organic contaminants during nanofiltration. *Environ. Sci. Technol.* 2010, *44* (7), 2592–2598.

[100] Braeken, L.; Ramaekers, R.; Zhang, Y.; Maes, G.; Bruggen, B. Van der; Vandecasteele, C. Influence of hydrophobicity on retention in nanofiltration of aqueous solutions containing organic compounds. *J. Memb. Sci.* 2005, *252* (1), 195–203.

[101] Shim, Y.; Rixey, W.; Chellam, S. Influence of sorption on removal of tryptophan and phenylalanine during nanofiltration. *J. Memb. Sci.* 2008, *323* (1), 99–104.

[102] Agenson, K. O.; Oh, J.-I.; Urase, T. Retention of a wide variety of organic pollutants by different nanofiltration/reverse osmosis membranes: controlling parameters of process. *J. Memb. Sci.* 2003, *225* (1), 91–103.

[103] Shim, Y.; Chellam, S. Steric and electrostatic interactions govern nanofiltration of amino acids. *Biotechnol. Bioeng.* 2007, *98* (2), 451–461.

[104] Verliefde, A. R. D.; Heijman, S. G. J.; Cornelissen, E. R.; Amy, G.; Van der Bruggen, B.; van Dijk, J. C. Influence of electrostatic interactions on the rejection with NF and assessment of the removal efficiency during NF/GAC treatment of pharmaceutically active compounds in surface water. *Water Res.* 2007, *41* (15), 3227–3240.

[105] Yangali-Quintanilla, V.; Sadmani, A.; McConville, M.; Kennedy, M.; Amy, G. Rejection of pharmaceutically active compounds and endocrine disrupting compounds by clean and fouled nanofiltration membranes. *Water Res.* 2009, *43* (9), 2349–2362.

[106] Verliefde, A. R. D.; Cornelissen, E. R.; Heijman, S. G. J.; Verberk, J. Q. J. C.; Amy, G. L.; Van der Bruggen, B.; van Dijk, J. C. The role of electrostatic interactions on the rejection of organic solutes in aqueous solutions with nanofiltration. *J. Memb. Sci.* 2008, *322* (1), 52–66.

[107] Derakhsheshpoor, R.; Homayoonfal, M.; Akbari, A.; Mehrnia, M. Amoxicillin separation from pharmaceutical wastewater by high permeability polysulfone nanofiltration membrane. *J. Environ. Heal. Sci. Eng.* 2013, *11* (1), 9.

[108] Bellona, C.; Drewes, J. E. The role of membrane surface charge and solute physico-chemical properties in the rejection of organic acids by NF membranes. *J. Memb. Sci.* 2005, *249* (1-2), 227–234.

[109] Nghiem, L. D.; Schäfer, A. I.; Elimelech, M. Role of electrostatic interactions in the retention of pharmaceutically active contaminants by a loose nanofiltration membrane. *J. Memb. Sci.* 2006, *286* (1-2), 52–59.

[110] Van der Bruggen, B.; Mänttäri, M.; Nyström, M. Drawbacks of applying nanofiltration and how to avoid them: A review. *Sep. Purif. Technol.* 2008, *63* (2), 251–263.

[111] Lau, W. J.; Ismail, A. F. Theoretical studies on the morphological and electrical properties of blended PES/SPEEK nanofiltration membranes using different sulfonation degree of SPEEK. *J. Memb. Sci.* 2009, *334* (1), 30–42.

[112] Chellam, S.; Taylor, J. S. Simplified Analysis of Contaminant Rejection During Ground- and Surface Water Nanofiltration Under the Information Collection Rule. *Water Res.* 2001, *35* (10), 2460–2474.

[113] Chang, E.-E.; Liang, C.-H.; Huang, C.-P.; Chiang, P.-C. A simplified method for elucidating the effect of size exclusion on nanofiltration membranes. *Sep. Purif. Technol.* 2012, *85*, 1–7.

[114] Chang, E.-E.; Chang, Y.-C.; Liang, C.-H.; Huang, C.-P.; Chiang, P.-C. Identifying the rejection mechanism for nanofiltration membranes fouled by humic acid and calcium ions exemplified by acetaminophen, sulfamethoxazole, and triclosan. *J. Hazard. Mater.* 2012, *221-222*, 19–27.

[115] Bowen, W. R.; Mukhtar, H. Characterisation and prediction of separation performance of nanofiltration membranes. *J. Memb. Sci.* 1996, *112* (2), 263–274.

[116] Schaep, J. Modelling the retention of ionic components for different nanofiltration membranes. *Sep. Purif. Technol.* 2001, *22-23* (1-2), 169–179.

[117] Bandini, S.; Vezzani, D. Nanofiltration modeling: the role of dielectric exclusion in membrane characterization. *Chem. Eng. Sci.* 2003, *58* (15), 3303–3326.

[118] Kotrappanavar, N. S.; Hussain, A. A.; Abashar, M. E. E.; Al-Mutaz, I. S.; Aminabhavi, T. M.; Nadagouda, M. N. Prediction of physical properties of nanofiltration membranes for neutral and charged solutes. *Desalination* 2011, *280* (1-3), 174–182.

[119] Oatley, D. L.; Llenas, L.; Pérez, R.; Williams, P. M.; Martínez-Lladó, X.; Rovira, M. Review of the dielectric properties of nanofiltration membranes and verification of the single oriented layer approximation. *Adv. Colloid Interface Sci.* 2012, *173*, 1–11.

[120] Geraldes, V.; Brites Alves, A. M. Computer program for simulation of mass transport in nanofiltration membranes. *J. Memb. Sci.* 2008, *321* (2), 172–182.

[121] Roy, Y.; Sharqawy, M. H.; Lienhard, J. H. Modeling of flat-sheet and spiral-wound nanofiltration configurations and its application in seawater nanofiltration. *J. Memb. Sci.* 2015, *493*, 360–372.

[122] Wu, F.; Feng, L.; Zhang, L. Rejection prediction of isopropylantipyrine and antipyrine by nanofiltration membranes based on the Spiegler–Kedem–Katchalsky model. *Desalination* 2015, *362*, 11–17.

[123] Lee, S.; Cho, J.; Elimelech, M. Combined influence of natural organic matter (NOM) and colloidal particles on nanofiltration membrane fouling. *J. Memb. Sci.* 2005, *262* (1), 27–41.

[124] Bartels, C.; Wilf, M.; Casey, W.; Campbell, J. New generation of low fouling nanofiltration membranes. *Desalination* 2008, *221* (1-3), 158–167.

[125] Nghiem, L. D.; Coleman, P. J.; Espendiller, C. Mechanisms underlying the effects of membrane fouling on the nanofiltration of trace organic contaminants. *Desalination* 2010, *250* (2), 682–687.

[126] Her, N.; Amy, G.; Chung, J.; Yoon, J.; Yoon, Y. Characterizing dissolved organic matter and evaluating associated nanofiltration membrane fouling. *Chemosphere* 2008, *70* (3), 495–502.

[127] Nghiem, L. D.; Hawkes, S. Effects of membrane fouling on the nanofiltration of pharmaceutically active compounds (PhACs): Mechanisms and role of membrane pore size. *Sep. Purif. Technol.* 2007, *57*, 176–184.

[128] Wei, X.; Wang, Z.; Fan, F.; Wang, J.; Wang, S. Advanced treatment of a complex pharmaceutical wastewater by nanofiltration: Membrane foulant identification and cleaning. *Desalination* 2010, *251* (1-3), 167–175.

[129] Contreras, A. E.; Kim, A.; Li, Q. Combined fouling of nanofiltration membranes: Mechanisms and effect of organic matter. *J. Memb. Sci.* 2009, *327* (1-2), 87–95.

[130] Feng, G.; Chu, H.; Dong, B. Characterizing dissolved organic matter fouling of nanofiltration membranes and evaluating effects of naproxen retention. *Desalin. Water Treat.* 2014, *56* (11), 1–13.

[131] Nghiem, L. D.; Hawkes, S. Effects of membrane fouling on the nanofiltration of trace organic contaminants. *Desalination* 2009, *236* (1-3), 273–281.

[132] Cobry, K. D.; Yuan, Z.; Gilron, J.; Bright, V. M.; Krantz, W. B.; Greenberg, A. R. Comprehensive experimental studies of early-stage membrane scaling during nanofiltration. *Desalination* 2011, *283*, 40–51.

[133] Tu, K. L.; Chivas, A. R.; Nghiem, L. D. Effects of membrane fouling and scaling on boron rejection by nanofiltration and reverse osmosis membranes. *Desalination* 2011, *279* (1), 269–277.

[134] Xu, P.; Bellona, C.; Drewes, J. E. Fouling of nanofiltration and reverse osmosis membranes during municipal wastewater reclamation: Membrane autopsy results from pilot-scale investigations. *J. Memb. Sci.* 2010, *353* (1), 111–121.

[135] Rana, D.; Scheier, B.; Narbaitz, R. M.; Matsuura, T.; Tabe, S.; Jasim, S. Y.; Khulbe, K. C. Comparison of cellulose acetate (CA) membrane and novel CA membranes containing surface modifying macromolecules to remove pharmaceutical and personal care product micropollutants from drinking water. *J. Memb. Sci.* 2012, *409-410*, 346–354.

[136] Narbaitz, R. M.; Rana, D.; Dang, H. T.; Morrissette, J.; Matsuura, T.; Jasim, S. Y.; Tabe, S.; Yang, P. Pharmaceutical and personal care products removal from drinking water by modified cellulose acetate membrane: Field testing. *Chem. Eng. J.* 2013, *225*, 848–856.

[137] Chaabane, T.; Zaidi, S.; Darchen, A.; Maachi, R. Treatment of pharmaceutical effluent by electrocoagulation coupled to nanofiltration. *Desalin. Water Treat.* 2013, *51* (25-27), 4987–4997.

[138] Ganiyu, S. O.; van Hullebusch, E. D.; Cretin, M.; Esposito, G.; Oturan, M. A. Coupling of membrane filtration and advanced oxidation processes for removal of pharmaceutical residues: A critical review. *Sep. Purif. Technol.* 2015, *156*, 891–914.

[139] Szép, A.; Kertész, S.; László, Z.; Szabó, G.; Hodúr, C. Advanced treatment of pharmaceutical wastewater by nanofiltration and ozonation. *Acta Tech. Corviniensis - Bull. Eng.* 2012, *5* (1), 25–27.

[140] Zhang, X.; Liu, Y.; Sun, C.; Ji, H.; Zhao, W.; Sun, S.; Zhao, C. Graphene oxide-based polymeric membranes for broad water pollutant removal. *RSC Adv.* 2015, *5* (122), 100651–100662.

[141] Esmaeili, A.; Entezari, M. H. Cubic Ag/AgBr–graphene oxide nanocomposite: sono-synthesis and use as a solar photocatalyst for the degradation of DCF as a pharmaceutical pollutant. *RSC Adv.* 2015, *5* (117), 97027–97035.

[142] García-Vaquero, N.; Lee, E.; Jiménez Castañeda, R.; Cho, J.; López-Ramírez, J. A. Comparison of drinking water pollutant removal using a nanofiltration pilot plant powered by renewable energy and a conventional treatment facility. *Desalination* 2014, *347*, 94–102.

[143] Liu, P.; Zhang, H.; Feng, Y.; Yang, F.; Zhang, J. Removal of trace antibiotics from wastewater: A systematic study of nanofiltration combined with ozone-based advanced oxidation processes. *Chem. Eng. J.* 2014, *240*, 211–220.

[144] Miralles-Cuevas, S.; Oller, I.; Pérez, J. A. S.; Malato, S. Application of solar photo-Fenton at circumneutral pH to nanofiltration concentrates for removal of pharmaceuticals in MWTP effluents. *Environ. Sci. Pollut. Res. Int.* 2015, *22* (2), 846–855.

[145] Taheran, M.; Brar, S. K.; Verma, M.; Surampalli, R. Y.; Zhang, T. C.; Valero, J. R. Membrane processes for removal of pharmaceutically active compounds (PhACs) from water and wastewaters. *Science of the Total Environment*. 2016.

[146] Bakangura, E.; Wu, L.; Ge, L.; Yang, Z.; Xu, T. Mixed matrix proton exchange membranes for fuel cells: State of the art and perspectives. *Prog. Polym. Sci.* 2016, *57*, 103–152.

[147] Hatakeyama, K.; Karim, M. R.; Ogata, C.; Tateishi, H.; Funatsu, A.; Taniguchi, T.; Koinuma, M.; Hayami, S.; Matsumoto, Y. Proton Conductivities of Graphene Oxide Nanosheets: Single, Multilayer, and Modified Nanosheets. *Angew. Chemie* 2014, *126* (27), 7117–7120.

[148] Karim, M. R.; Hatakeyama, K.; Matsui, T.; Takehira, H.; Taniguchi, T.; Koinuma, M.; Matsumoto, Y.; Akutagawa, T.; Nakamura, T.; Noro, S.; Yamada, T.; Kitigama, H.; Hayami, S. Graphene Oxide Nanosheet with High Proton Conductivity. 2013.

[149] Gao, W.; Wu, G.; Janicke, M. T.; Cullen, D. A.; Mukundan, R.; Baldwin, J. K.; Brosha, E. L.; Galande, C.; Ajayan, P. M.; More, K. L.; Dettelbaum, A. L.; Zelenay, P. Ozonated Graphene Oxide Film as a Proton-Exchange Membrane. *Angew. Chemie Int. Ed.* 2014, *53* (14), 3588–3593.

[150] Chu, F.; Lin, B.; Feng, T.; Wang, C.; Zhang, S.; Yuan, N.; Liu, Z.; Ding, J. Zwitterion-coated graphene-oxide-doped composite membranes for proton exchange membrane applications. *J. Memb. Sci.* 2015, *496*, 31–38.

[151] Paneri, A.; Heo, Y.; Ehlert, G.; Cottrill, A.; Sodano, H.; Pintauro, P.; Moghaddam, S. Proton selective ionic graphene-based membrane for high concentration direct methanol fuel cells. *J. Memb. Sci.* 2014, *467*, 217–225.

[152] Peng, K.-J.; Lai, J.-Y.; Liu, Y.-L. Nanohybrids of graphene oxide chemically-bonded with Nafion: Preparation and application for proton exchange membrane fuel cells. *J. Memb. Sci.* 2016, *514*, 86–94.

In: Nanosheets and Nanospheres
Editor: Deborah Wright
ISBN: 978-1-53610-203-1

Chapter 2

TRANSITION METAL DICHALCOGENIDE NANOSHEETS FOR SENSOR PLATFORMS

Yong Wang*[*] *and Qianfen Zhuang
College of Chemistry, Nanchang University, Nanchang, China

ABSTRACT

In the past few years, transition metal dichalcogenide (TMD) nanosheets have aroused tremendous attention within the scientific community due to a wealth of unique new physics and chemistry that emerges as its thickness goes down to single or few layer. The unusual and fascinating physicochemical properties of these TMD nanosheets make them intriguing candidates for use in optoelectronic devices, energy storage devices and catalysis. However, in very recent times, there was a significant emerging trend in their utilization in sensors. Therefore, in this book chapter, we attach great importance to this new trend for the sensing applications, and summarize some representative works of applications of TMD nanosheets in sensing area. Moreover, current challenges and future perspectives in the sensing area are also outlined.

Keywords: transition metal dichalcogenide, nanosheet, sensors

* College of Chemistry, Nanchang University, Nanchang, China. Corresponding Author: Yong Wang, Email: wangyong@ncu.edu.cn.

1. INTRODUCTION

First isolated in 2004 by the physicists Andre Geim and Konstantin Novoselov at the University of Manchester, graphene has since captured extensive global interests across academia and industry, particularly following Geim and Novoselov being awarded the 2010 Nobel Prize in Physics for "groundbreaking experiments regarding the two-dimensional (2D) material graphene" [1-8]. This huge interest arises from the material's unprecedented physical and chemical properties, such as excellent mechanical flexibility, large surface area, unparalleled electrical conductivity, extraordinary thermal conductivity, superior catalytic activity, remarkable biocompatibility and ease of functionalization [9-13]. Benefiting from the exceptional properties, graphene exhibits a myriad of promising applications in the fields of supercapacitors, catalysis, drug delivery, sensors and energy storage [9-13]. The unique and unusual physicochemical properties of graphene stem from its ultra-thin planar structure and two dimensional morphology to a large degree [14, 15]. Therefore, great attention has been recently devoted to other inorganic grapheme analogous two-dimensional layered materials, especially those ultrathin transition metal dichalcogenide (TMD) nanosheets [16-18].

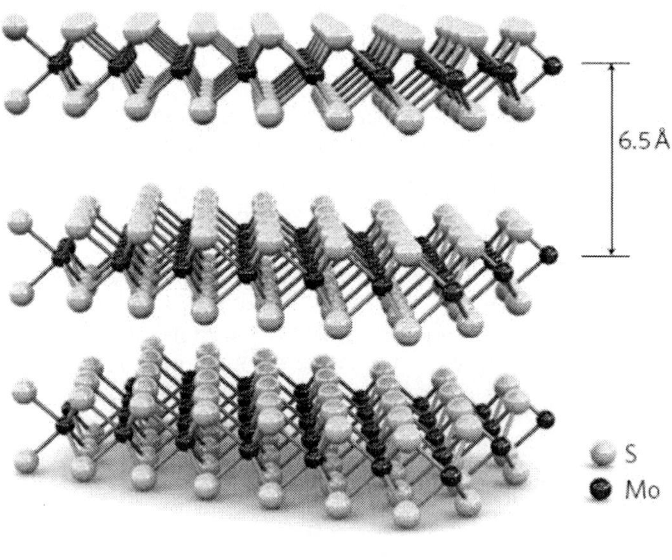

Figure 1. Three-dimensional model of the crystal structure of TMD with a general formula of MX_2. Adapted with permission from ref. 28. Copyright 2011 Macmillan Publishers Ltd.

TMD, whose general molecular formula is represented as MX_2 (M denotes a transition metal of group 4-10 and X denotes a chalcogen), possesses a graphene-like layered structure containing one atomic layer of metal atoms with 6-fold in-plane coordination, hexagonally packed between two trigonal atomic layers of chalcogenide atoms (see Figure 1) [17, 18]. The triple layers are stacked and held together by interlayer van der Waals forces to form a layered structure of the MX_2 crystal. It is noteworthy that since TMD shows a similar two-dimensional lamellar structure to graphene, it also has extraordinary physical and chemical properties, which was superior to or comparable with graphene [18-20]. In the last few years, the fascinating physicochemical properties of TMD nanomaterials have made them intriguing candidates for use in optoelectronic devices, energy storage devices and catalysis [21-28]. However, in very recent years, there was a significant emerging trend in their utilization in sensors [29-31]. A sensor is a material or device that can be used for the detection of a particular chemical species of interest at trace levels. Sensing-based detection approach is usually superior to traditional detection methods including mass spectrometry, chromatography and titration, in terms of speediness of analysis, low cost of instrumentation and operation, simplicity for operators, suitability for mass sample selection and on-site detection. Hence, on the basis of its inherent advantages, sensing-based detection method has been currently considered to be the preferred choice for the detection of trace amounts of specific analyte, and has found a broad range of applications, such as environmental and industrial monitoring, drug quality monitoring, point-of-care diagnostics, forensic analysis, food safety, as well as national security and defense [32-34]. Depending on the transduction mechanism and readout method, sensors can be respectively classified into four groups: optical, electrochemical, field-effect transistor-based, and mass-sensitive. In this chapter, we focus on the applications of TMD nanosheets in devising sensors for detection of analyte of interest, and summarize some representative works of applications of TMD nanosheets in the aforementioned four type of sensors.

2. Applications of TMD Nanosheets for Sensing

2.1. Optical Sensors

Optical sensors are pretty attractive because they are simple, fast, sensitive, inexpensive, non-destructive, and can be well-suited for visual detection by the naked eye. On the basis of various optical principles, optical sensors can be generally categorized into four types: colorimetric sensors, fluorescence sensors, chemiluminescence sensors, and surface plasmon resonance sensors.

2.1.1. Colorimetric Sensors

In 2007, Professor Xiyun Yan and her colleagues from Institute of Biophysics in China made the surprising discovery that magnetite nanoparticles actually showed an intrinsic peroxidase-like activity [35]. Since then, many researchers started to explore the peroxidase-like property of other nanomaterials such as gold nanoparticles (AuNPs), cerium oxide nanoparticles, graphene nanosheets, carbon dots, and so on, and found that the nanozymes possessed many advantages in several aspects including low cost, ease of mass production, robustness to harsh environments, and high stability [36-38].

In 2014, Lin et al. gave the first demonstration of the peroxidase-like activity of molybdenum disulfide (MoS_2) nanosheets which was prepared by solution exfoliation method, and found that MoS_2 nanosheets could catalytically oxidize peroxidase substrate 3,3,'5,5' -tetramethylbenzidine (TMB) by hydrogen peroxide (H_2O_2) to produce a blue-colored reaction. Depending on the catalytic activity of MoS_2 nanosheets and glucose oxidase, the phenomenon was exploited to fabricate a colorimetric sensor for detection of H_2O_2 and glucose (see Figure 2) [39]. Subsequently, Lin et al. [40] and Chen et al. [41] found respectively that tungsten disulfide (WS_2) nanosheets and hemin-functionalized WS_2 nanosheets had the similar peroxidase-like activity to MoS_2 nanosheets, and could be used for the colorimetric sensing of H_2O_2 and glucose. And Guo et al. [42] and Zhao et al. [43] found respectively that the MoS_2 nanosheets and surfactants modified MoS_2 nanoparticles, which were synthesized hydrothermally, also could catalyze the oxidation of TMB in the presence of H_2O_2 to produce a blue product, confirming that the MoS_2 nanosheets synthesized under hydrothermal conditions had the same intrinsic peroxidase-like activity as the MoS_2 nanosheets prepared by solution exfoliation method.

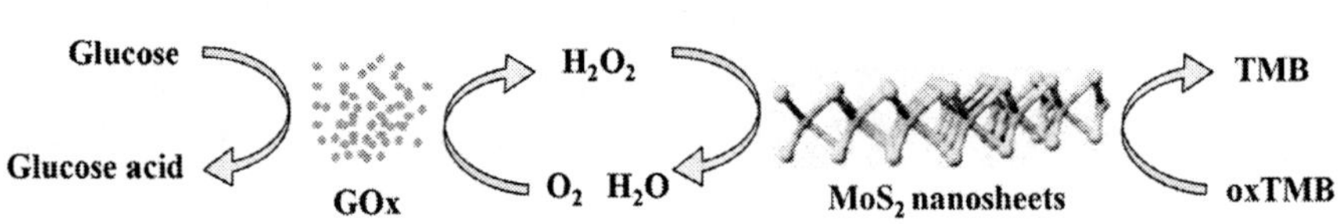

Figure 2. Schematic illustration of colorimetric sensing of H_2O_2 and glucose by using glucose oxidase (GOx) and MoS_2 nanosheet-catalyzed reactions. Adapted with permission from ref. 39. Copyright 2014 Royal Society of Chemistry.

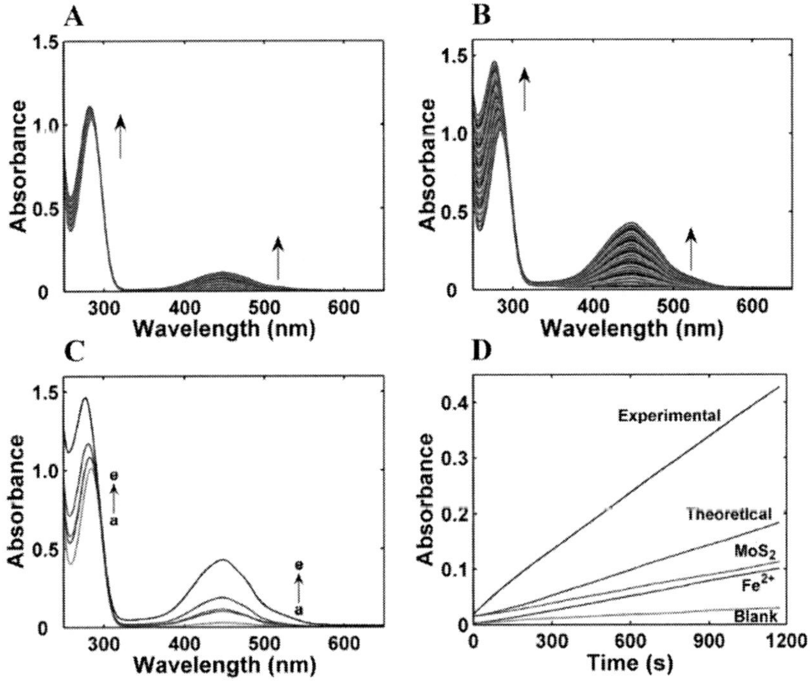

Figure 3. Time-dependent UV-Vis spectra between 0 s and 1172 s for (A) the $MoS_2/OPD/H_2O_2$ reaction, and (B) the $Fe^{2+}/MoS_2/OPD/H_2O_2$ reaction. (C) Absorption spectra at $t = 1172$ s. (a) blank, (b) $Fe^{2+}/OPD/H_2O_2$, (c) $MoS_2/OPD/H_2O_2$, (d) $Fe^{2+}/MoS_2/OPD/H_2O_2$ (theroretical), (e) $Fe^{2+}/MoS_2/OPD/H_2O_2$ (experimental). (e) $Fe^{2+}/MoS_2/OPD/H_2O_2$ (experimental). (D) Kinetic profiles at 446 nm for the OPD/H_2O_2 system (blank), the $MoS_2/OPD/H_2O_2$ system, the $Fe^{2+}/OPD/H_2O_2$ system, as well as the theoretical and experimental profiles of the $MoS_2/OPD/H_2O_2$ system and the $Fe^{2+}/OPD/H_2O_2$ system. Adapted with permission from ref. 47. Copyright 2016 Elsevier.

These works provoked an immediate increase in further research of these nanosheets and their applications in the area of sensing. Lei et al. [44] have recently reported that algae-like molybdenum disulfide/polypyrrole (MoS_2/PPy) nanocomposite through a redox reaction between ammonium tetrathiomolybdate and pyrrole monomer under a hydrothermal condition gave superior peroxidase-like catalytic activity compared with that of the MoS_2 or PPy alone, which is well-suited for colorimetric sensing of H_2O_2 with a high sensitivity. Nirala et al. [45] reported that the AuNPs when supported on molybdenum disulfide nanoribbons matrix presented synergistically enhanced intrinsic peroxidase-like catalytic activity, and took advantage of the synergistically catalytic activity for colorimetric detection of cholesterol. Cai et al. [46] reported that the MoS_2 nanosheets decorated with octahedral $Pt_{74}Ag_{26}$ alloy nanoparticles could greatly improve the catalytic activity, which could be used for highly sensitive colorimetric detection of H_2O_2. Very recently, we prepared luminescent MoS_2 nanosheets under hydrothermal conditions by using thiourea and sodium molybdate as precursors, and found that the luminescent MoS_2 nanosheets also had intrinsic peroxidase-mimicking catalytic activity, and efficiently catalyzed the oxidation of colorless substrate, o-phenylenediamine (OPD), by H_2O_2 to yield a color reaction [47]. When Fe^{2+} was added into the system, we found a remarkable increase in the catalytic activity relative to that of that of the luminescent MoS_2 nanosheets or the catalyst Fe^{2+} (see Figure 3). This synergistically enhanced catalytic property was then employed to construct a colorimetric catalytic sensor for detection of catalyst Fe^{2+}. It was demonstrated that the use of the luminescent MoS_2 nanosheets can greatly increase the sensitivity and selectivity of the Fe^{2+} sensor, and facilitated the visual detection of Fe^{2+}. More importantly, we submitted the three-way kinetic-spectral data of the sensing system to an advanced chemometrics method, multivariate curve resolution by alternating least squares (MCR-ALS), and found that the increased sensitivity and selectivity of the sensor was attributed to its large reaction rate constant.

2.1.2. Fluorescence Sensors

Currently, the development of MoS_2 nanosheet-based fluorescence sensors mainly depends on the mechanism of fluorescence resonance energy transfer (FRET) or charge transfer doping (CTD).

Fluorescence resonance energy transfer (FRET) involves the transfer of the excitation energy from a donor flurophore in an excitated electronic state to a neighboring acceptor molecule in a non-radiative manner through a dipole-dipole interaction [48]. Its occurrence requires a pair of appropriate energy donor and energy acceptor which must be in close proximity (typically 1.0-10.0 nm), and have spectral overlap (the absorption spectrum of the acceptor and the fluorescence emission spectrum of the donor). The FRET efficiency (E) depends on the donor-to-acceptor separation distance (d) with an inverse sixth power law: $E = 1/[1+(d/d_0)^6]$, where d_0 denotes a Föster radius, which is the distance between the donor and the acceptor at which the energy transfer efficiency is 50%. Depending on the FRET mechanism, many researchers exploited organic dye molecules as both the donor flurophore and the acceptor to develop various fluorescence sensors [49, 50]. However, the distance between dye molecules is not usually so large that the energy-transfer efficiency between the two is low. This unfavorable factor results in a low signal-to-noise ratio, which seriously affect the detection sensitivity of analyte of interest. To this end, various kinds of nanomaterials such as nanoparticles, quantum dots, nanotubes, nanosheets, etc., as both energy donors as well as acceptors have quickly filled in the role, being found to be superior to traditional organic dyes on aforementioned count [51-53]. Many researchers found an d^{-4} dependence of the FRET efficiency (E) corresponding to nanomaterials on the separation distance between the donor and the acceptor [54-58]. The results suggest that compared with traditional FRET, FRET corresponding to nanomaterials is highly efficient in terms of energy-transfer efficiency. It is worth noting that FRET related to nanomaterials is still significant even when the distance is as long as 30 nm, breaking the traditional distance limitation of FRET which is no more than 10 nm. At present, researchers have found that MoS_2 nanosheets as a novel 2D nanomaterial could be used as highly efficient long-range fluorescence acceptors due to their superquenching efficiency in a wide spectral range. Therefore, the nanosheets have been applied for construction of FRET-based fluorescence sensors for detection of various substances, such as nucleic acids, proteins, metal ions, and small molecules.

In 2013, Zhang's group described the binding and quenching abilities of MoS_2 nanosheets against a fluorophore-labeled DNA for the first time [59]. As showed in Figure 4A, a dye-labeled single-stranded DNA (ssDNA) probe was first absorbed onto the surface of MoS_2 nanosheets via the van der Waals interaction between nucleobases and the basal plane of MoS_2. This caused almost entirely quenching of the dye fluorescence due to the FRET interaction between the acceptor of MoS_2 nanosheets and the dye donor. Upon the addition of the complementary target DNA, the hybridization between probe DNA and target DNA weakened the interaction of nucleobases with the MoS_2 nanosheets, leading to the desorption of probe DNA and recovery of the dye fluorescence. As a result, the recovered fluorescence of the DNA probe could be used for a quantitative readout of the target DNA. Besides the DNA detection, the group and Ge et al. further exploited the concept to construct an aptamer-based sensor for detection of adenosine and adenosine triphosphate (ATP) [59, 60]. As displayed in Figure 4B, the anti-adenosine aptamer or anti-ATP aptamer labeled with a dye molecule was physically adsorbed onto the MoS_2 nanosheets. The addition of adenosine resulted in the conformation change of anti-adenosine aptamer from an unfolded and flexible structure to a rigid and definite tertiary structure. The anti-adenosine aptamer with rigid structure had weak affinity toward the MoS_2 nanosheets, leading to the restoration of the dye fluorescence.

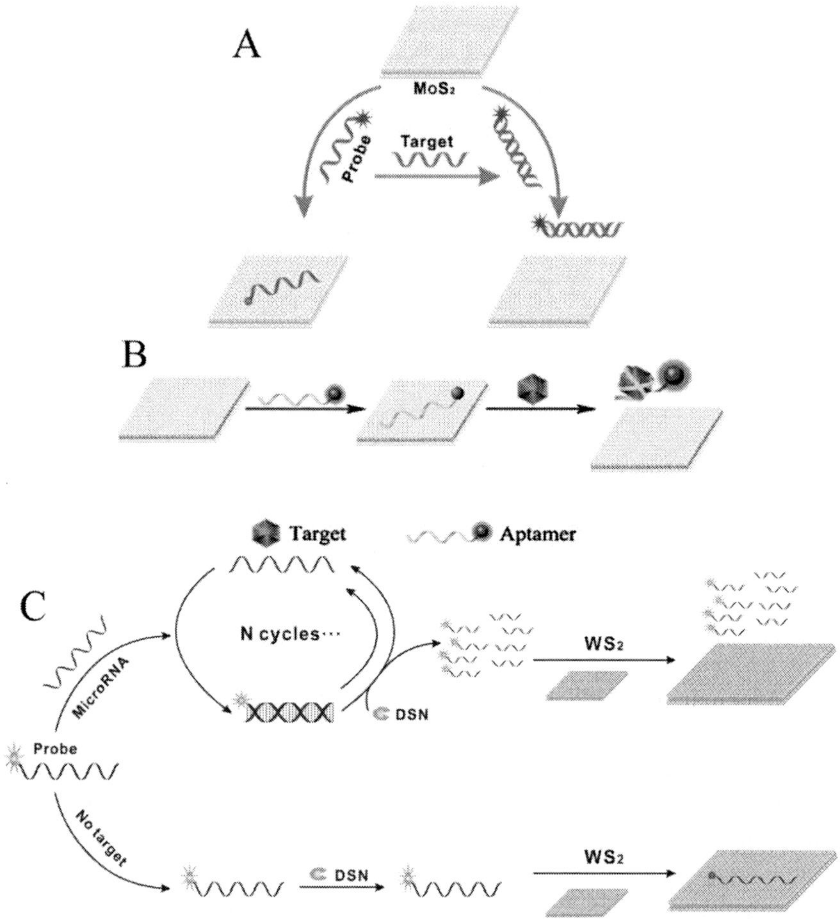

Figure 4. (A) Schematic illustration of MoS_2 nanosheet-based fluorometric sensing of DNA. Adapted with permission from ref. 59. Copyright 2013 American Chemical Society. (B) Schematic illustration of fluorescence sensing of target molecule (adenosine or adenosine triphosphate) based on the aptamer functionalized MoS_2 nanosheets. Adapted with permission from ref. 60. Copyright 2014 Royal Society of Chemistry. (C) Schematic illustration of WS_2 nanosheet–based sensing platform coupled with duplex-specific nuclease signal amplification for detection of microRNA. Copyright 2014 Royal Society of Chemistry. Adapted with permission from ref. 74. Copyright 2014 American Chemical Society.

Combined with the use of assorted specific functional DNA (aptamers or DNAzymes), similar sensors have also been fabricated by other groups to detect various targets, including human α-thrombin [60], prostate specific antigen

[62], silver ions [63], and uranyl ions [64]. Moreover, the sensor strategy based on MoS_2 nanosheets can be further integrated into a microfluidic device for rapid detection of DNA [65]. After that, inspired by excellent performance of the sensor based on MoS_2 nanosheets, many researchers synthesized other 2D TMD nanosheets like WS_2, VS_2, Ta_2NiS_5, TaS_2 and TiS_2, and found that similar to MoS_2 nanosheets, these nanomaterials also possessed high quenching efficiency, and could discriminate unfolded ssDNA and dsDNA/rigid ssDNA [66-69]. Then, these researchers used these TMD nanosheets for sensing of DNA and carcinoembryonic antigen. Among them, Zhang et al. further exploited the above-mentioned sensing strategy to achieve the multiplexed fluorescence detection of DNA in the presence of TaS_2 nanosheets [69]. Owing to the large planar surface, TaS_2 nanosheets allows simultaneous adsorbing and quenching of two different dye-labeled ssDNA probes which target Influenza A virus subtype H1N1 gene and subtype H5N1 gene. After the addition of the two specific target sequence, the dye-labeled ssDNA probes can recover the corresponding fluorescence. Lately, Wang et al. employed a peptide nucleic acid (PNA) probe instead of a DNA probe for the first time, and adopted a similar sensing strategy to achieve target DNA detection in the presence of WS_2 nanosheets [70]. Because the binding affinity and sequence specificity of the PNA probe to the target DNA are superior to those of the DNA probe, and the related hybridization product, PNA-DNA, has a higher stability than DNA-DNA, the use of PNA probe can improve the performance of the DNA hybridization sensor. In addition, benefiting from the difference in the affinity of TMD nanosheets toward ssDNA and dsDNA, Ge et al. exploited the phosphorylation-specific exonuclease reaction and fluorescence quenching of dye-labeled ssDNA by WS_2 nanosheets to develop the sensors for detection of T4 polynucleotide kinase and its inhibitors [71]. Huang et al. combined the hybridization chain reactions with the fluorescence quenching of dye labeled ssDNA by MoS_2 nanosheets to amplify the fluorescence signal and reduce the background signal, and finally achieved the detection of target DNA [72]. Deng et al. took full advantage of the property of DNA hybridization, DNA adenosine methyltransferase and DpnI endonuclease to design a MoS_2 nanosheet-based fluorescence sensor for detection of DNA methyltransferase activity [73]. More importantly, Xi et al. found unexpectedly that WS_2 nanosheets possessed a capability for discriminating short DNA fragment (<10 bases) and ssDNA probe (>10 bases), and exploited the finding to design a WS_2 nanosheet-based sensing platform coupled with duplex-specific nuclease signal amplification for detection of microRNA [74]. As displayed in Figure 4C, the dye-labeled ssDNA probe hybridizes to microRNA to give a DNA/RNA heteroduplex. The addition

of duplex-specific nuclease (DSN) can cleave the heteroduplex into short DNA fragment and release the microRNA. The released microRNA can hybridize with another ssDNA probe, which initiates a next round of cleavage, releasing, and hybridization. The combination of the cyclic reaction and the discrimination ability of WS_2 nanosheets for short DNA fragment and ssDNA probe can amplify the fluorescence signal of microRNA detection. Then, the aforementioned finding is further expanded for fluorescence sensing of nuclease and bleomycin detection on the basis of nuclease-catalyzed hydrolysis of ssDNA or bleomycin/Fe^{2+}–induced oxidation cleavage [75].

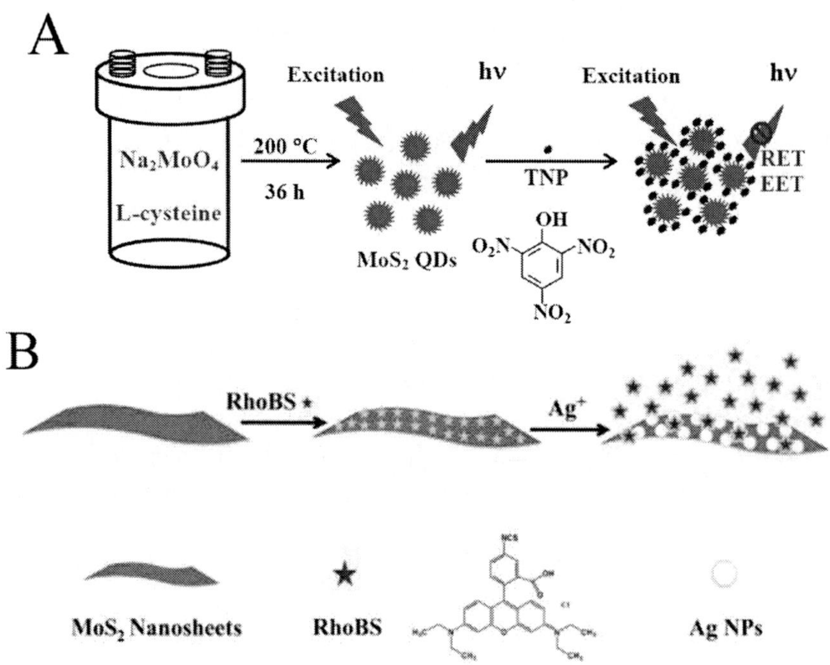

Figure 5. (A) Schematic illustration of MoS_2 QD-based fluorescence quenching sensing of TNP. Adapted with permission from ref. 76. Copyright 2014 American Chemical Society. (B) Schematic illustration of fluorescence sensing of Ag^+ based on rhodamine B isothiocyanate–adsorbed MoS_2 nanosheets. Adapted with permission from ref. 77. Copyright 2015 American Chemical Society.

In 2014, our group has synthesized fluorescent MoS_2 quantum dots (QDs) by using sodium molybdate and cysteine as precursors, and applied the MoS_2 QDs to fabricate a fluorescence quenching sensor for detection of 2,4,6-

trinitrophenol (TNP) (see Figure 5A) [76]. It was found that the sensor exhibited displayed a high sensitivity toward TNP over other structurally similar substances including 2,4,6-trinitrotoluene, p-chlorophenol, phenol, and 2,6-di-tert-butyl-4-methylphenol. The high sensitivity toward TNP was demonstrated to be ascribed to a combination of the fluorescence resonance energy transfer, electronic energy transfer, and electrostatic interactions between MoS_2 QDs and TNP by using the proposed multiwavelength Stem-Volmer equation. Furthermore, Yang et al. constructed a fluorescence sensor based on rhodamine B isothiocyanate-adsorbed MoS_2 nanosheets for Ag^+ detection (see Figure 5B) [77]. The addition of MoS_2 nanosheets into rhodamine B isothiocyanate resulted in the fluorescence quenching of the dye via FRET. Further addition of Ag^+ remarkably enhanced the fluorescence of the system. This was because Ag^+ can be reduced to Ag nanoparticles on the surface of MoS_2 nanosheets, leading to the release of rhodamine B isothiocyanate molecules and surface-enhanced fluorescence from rhodamine B isothiocyanate remaining adsorbed on MoS_2.

In the past few years, many researchers have found that TMD nanosheets not only acted as a fluorescence quencher, but also had intrinsic fluorescence [78, 79]. The fluorescence modulation (enhancement or diminishiment) of TMD nanosheets was reported to be dominated by charge transfer mechanism between dopant molecules and TMD nanosheets, i.e., the fluorescence intensity of TMD nanosheets was related to their electron density. Generally, the increase in the total electron density of the *n*-type TMD nanosheets causes their fluorescence diminishiment, or vice versa. On the basis of the charge transfer doping (CTD) mechanism, Laon et al. devised a graphene/MoS_2 heterostructure by stacking graphene on a MoS_2 monolayer nanosheet and used the heterostructure to construct a sensor for DNA detection [78]. The introduction of the graphene could protect MoS_2 layer from the effect of the ambient environment as well as a biocompatible interface, and thus the graphene was well-suited for hosting DNA molecules on its surface. It was found that the immobilization of probe DNA onto the surface of the heterostructure had almost no influence on the fluorescence of MoS_2 nanosheets, and the further addition of complementary DNA led to the significant fluorescence enhancement due to the CTD effect. Moreover, our group prepared fluorescent MoS_2 nanosheets under hydrothermal conditions by employing sodium molybdate and thiourea as the starting materials, and found that the addition of Pb^{2+} induced the fluorescence enhancement of MoS_2 nanosheets, and the successive addition of S^{2-} led to the fluorescence quenching of the nanosheets [79]. The results were ascribed to the CTD mechanism between these ions and MoS_2 nanosheets. As can be seen from Figure 6A, we exploited the above-mentioned observations to fabricate a label-free "turn on-off" fluorescent nanosensor for detection of Pb^{2+} and S^{2-}. In addition to the above-mentioned fluorescence sensors based on FRET and CTD mechanisms, our group combined the peroxidase-like activity of MoS_2 nanosheets with fluorescence technique to fabricate a novel label-free fluorescent catalytic sensor for Fe^{2+} detection (see Figure 6B) [80]. However, the fluorescence sensor was not sufficient for quantitative detection of Fe^{2+} in environmental water samples because the water samples usually contained the natural interferents, leading to a matrix effect or/and unknown fluorescent substances. To this end, we used an advanced chemometrics method based on parallel factor analysis (PARAFAC) to deal with the excitation-emission matrix data related to the fluorescence spectra of the real water samples. The results showed that the introduction of the chemometrics approach could effectively eliminate the interferents from real water samples.

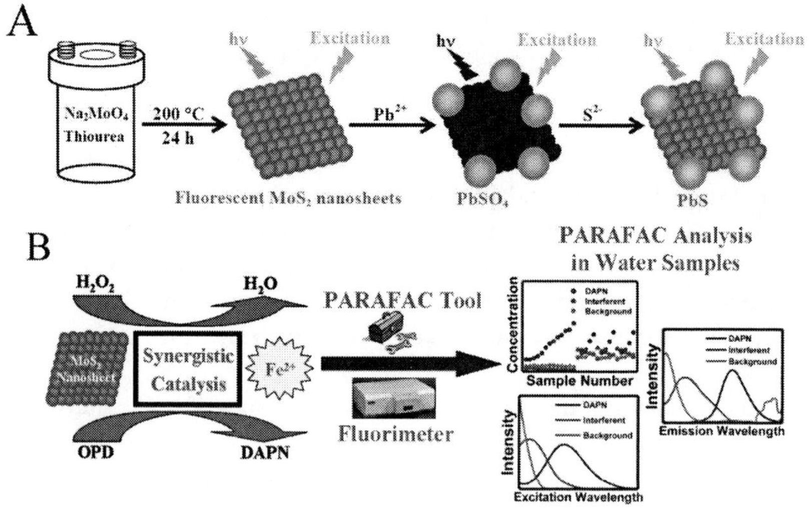

Figure 6. (A) Schematic illustration of label-free "turn on-off" fluorescent sensor of Pb^{2+} and S^{2-} based fluorescent MoS_2 nanosheets. Adapted with permission from ref. 79. Copyright 2016 American Chemical Society. (B) Schematic illustration of label-free fluorescent catalytic sensor of Fe^{2+} detection in water samples using MoS_2 nanosheets coupled with an advanced chemometric method. Adapted with permission from ref. 80. Copyright 2016 Royal Society of Chemistry.

2.1.3. Other Optical Sensors

Besides colorimetric sensors and fluorescence sensors, other TMD nanosheet-based optical sensors like chemiluminescence sensors and surface plasmon resonance (SPR) sensors have been recently studied by researchers [81, 82]. Zhao et al. developed a WS_2 nanosheet-based chemiluminescence resonance energy transfer (CRET) platform for sensing microRNA via the CRET process between the chemiluminescent donor luminol and chemiluminescent acceptor WS_2 nanosheets [81]. The ssDNA containing the G-quadruplex-hemin DNAzyme structure was adsorbed onto the surface of WS_2 nanosheets, leading to the chemiluminescence quenching of luminol due to CRET effect. When microRNA was added into the system, it partly hybridized with ssDNA to give a stable DNA-RNA heteroduplex. The heteroduplex could not be effectively adsorbed onto the surface of WS_2 nanosheets, resulting in the enhancement of chemiluminescence. Zeng et al. designed a new SPR sensor based on graphene/MoS_2 heterostructure for detection of biomolecules [82]. The addition of MoS_2 nanosheets in the system could remarkably increase the sensitivity of the SPR due to the excellent light absorption rate of MoS_2 nanosheets.

2.2. Electrochemical Sensors

Electrochemical techniques show great potential in the field of sensing due to the advantages of simplicity, cheapness, on-site monitoring, high sensitivity and high selectivity. Recently, TMD nanosheets have been widely used as the sensing materials for electrochemical sensors because they possess high electrical conductivity and many activity sites, and offer a platform for the electronic activity part such noble metal nanoparticles, enzymes, antigens, and so on. Current sensors based on TMD nanosheets can be classified as amperometric sensors, impedimetric sensors, electrochemiluminescence sensors, and photoelectrochemical sensors, which are discussed as follows.

2.2.1. Amperometric Sensors

The amperometric sensors are the most successfully commercialized devices among number types of sensors. Therefore, many researchers have recently developed a variety of amperometric sensors based on TMD nanosheets.

In 2013, Wang et al. found the strong electrocatalytic activity of MoS_2 nanoparticles prepared by simple ultrasonication method toward the reduction of H_2O_2, and exploited this property to construct a non-enzymatic amperometric sensor for H_2O_2 detection (see Figure 7) [83]. At the same time, they immobilized the glucose oxidase onto MoS_2 nanoparticles for electrochemical detection of glucose. Then, researchers found that the nanocomposites of TMD nanosheets with other materials could improve the performance of electrochemical sensors owing to the synergistic effects. Huang et al. investigated the electrochemical performance of MoS_2-graphene modified electrode, and found that MoS_2-graphene composite film facilitated the electron transfer kinetics [84]. They employed the modified electrode to construct an amperometric sensor of acetaminophen detection. After that, Huang et al. found that WS_2-graphene composite film could also accelerate the electron transfer kinetics, and used the nanocomposite to fabricate an amperometric sensor for simultaneous detection of catechol, resorcinol and hydroquinone [85]. Jin's group employed chemical reduction method to prepare MoS_2 nanosheets decorated with copper nanoparticles, and found that this nanocomposite showed enhanced electrocatalytic activity toward the oxidation of glucose. Benefiting from this observation, they developed an enzymeless amperometric sensor for glucose detection [86]. Su et al. explored the AuNPs-decorated MoS_2 nanocomposite prepared by microwave-assisted approach to develop an enzymeless amperometric sensor for glucose detection [87]. Huang et al. employed MoS_2 nanoflower and chitosan-AuNPs composites modified glassy carbon electrode (GCE) to construct an amperometric sensor for detection of bisphenol A [88], and they constructed an amperometric sensor for dopamine detection based on MoS_2-polyaniline and AuNPs modified GCE [89]. Wang's group synthesized AuNPs@MoS_2 nanocomposite by electrodeposition of AuNPs on the MoS_2 nanosheets, and used the material to develop an amperometric sensor for detection of dopamine, uric acid and ascorbic acid [90]. Li's group used Ni nanoparticle-MoS_2 nanocomposite prepared by chemical reducing Ni nanoparticles on the MoS_2 nanosheets for enzymeless amperometric sensing of gluocose [91]. Zhao et al. exploited layered MoS_2/graphene nanocomposites synthesized hydrothermally to develop a honokiol amperometric sensor [92]. Feng et al. synthesized hydrothermally poly(diallyldimethylammonium chloride) functionalized MoS_2-graphene nanoflower materials (PDDA-MoS_2-graphene) and assembled AuNPs on the surface of the material via electrostatic attraction to form AuNPs/PDDA-MoS_2-graphene nanocomposite, and then exploited the nanocomposite to construct a amperometric sensor for detection of eugenol [93]. Yang et al. placed the MoS_2-polyaniline prepared via ultrasonic exfoliation method on GCE to construct an amperometric sensor for detection of bisphenol A or chloramphenicol [94, 95]. Zheng et al. presented a facile process of the hydrothermal preparation of MoS_2 nanocube/poly(diallyl dimethyl ammonium chloride)-mesoporous carbon (MoS_2/PDDA-MC), and found that the GCE modified with MoS_2/PDDA-MC possessed catalytic activity toward L-cysteine oxidation. On the basis of the observation, they construct an amperometric sensor for L-cysteine detection [96]. Lin et al. reported the use of GCE modified with a hybrid material consisting of multiwalled carbon nanotubes and MoS_2 to fabricate a non-enzymatic amperometric sensor for H_2O_2 detection [97]. Kim et al. presented a facile process of *in situ* synthesis of MoS_2 nanosheets on the Au electrode, then immobilized horseradish peroxidase (HRP) conjugated IgG via the electrostatic attraction to form and Au-MoS_2/HRP hybrid, and finally exploited the hybrid material to construct an amperometric sensor for H_2O_2 detection [98]. Chao et al. immobilized hemoglobin on a AuNPs-decorated MoS_2 nanosheet nanocomposite film modified GCE, and found that the modified electrode had good electrocatalytic activity toward the H_2O_2 or NO reduction. Therefore, they used the nanocomposite as a platform to fabricate an amperometric sensor for H_2O_2 or NO detection [99]. Pramoda et al. used MoS_2-reduced graphene oxide composite as electrode for amperometric sensing of dopamine and uric acid in the presence of ascorbic acid [100]. Yang et al. employed MoS_2-polyaniline composite prepared by ultrasonic exfoliating method as electrode for amperometric sensing of adenine and guanine [101]. Zhao's group adopted two-step method to synthesize three-dimensional porous H_xTiS_2 nanosheet-polyaniline nanocomposite, and took advantage of the nanocomposite to develop highly selective amperometric sensors for Cu^{2+} detection. It was found that the high selectivity of the sensor for Cu^{2+} originated from coordination interaction between Cu^{2+} and the N atoms of the imine moieties in polyaniline [102]. Chao et al. prepared platinum nanoparticles-decorated MoS_2 nanocomposite by microwave-assisted hydrothermal method, and used the material as electrode for amperometric sensing of dopamine and uric acid [103]. Zhang et al. described the use of Fe_3O_4/MoS_2 nanocomposite prepared under hydrothermal conditions to construct an amperometric sensor for detection of Cr(VI) and nitrite [104]. Vasilescu et al. exploited the MoS_2-graphene quantum dot nanocomposite as a support for laccase immobilization, and modified them on the surface of carbon screen–printed electrode. Then, they used the enzyme electrode for amperometric sensing of caffeic acid, chlorogenic acid, and polyphenolic compounds [105]. Zhu et al. prepared PtW/MoS_2 hybrid nanocomposite through in-situ growth of PtW nanocrystals on the surface of MoS_2 nanosheets, and employed the material for enzymeless amperometric sensing of H_2O_2 released from living cells [106].

Owing to the naturally semiconducting properties of bulk MoS_2, the conductive ability of the aforementioned MoS_2 nanomaterials is usually limited. To this end, Professor Zhang from Nanyang Technological University have recently found that MoS_2 nanosheets attached to the glassy carbon electrode could be electrochemically reduced to produce the reduced MoS_2 ($rMoS_2$) with good conductivity and fast electron transfer rate. It was reported that the reduced MoS_2 was a good electrode material for amperometric sensing applications, such as sensing of dopamine, ascorbic acid and uric acid [107]. After that, our group attempted to use 32-mer homoadenine ssDNA oligonucleotides (A_{32}) and graphene as an effective anchoring block for binding to the MoS_2 nanosheet and the Au electrode, and then the binding MoS_2 on the surface of Au electrode was electrochemically reduced to produce the $rMoS_2$ (see Figure 8A) [108]. The modified $rMoS_2$-graphene/A_{32}/Au electrode was used for amperometric sensing of riboflavin.

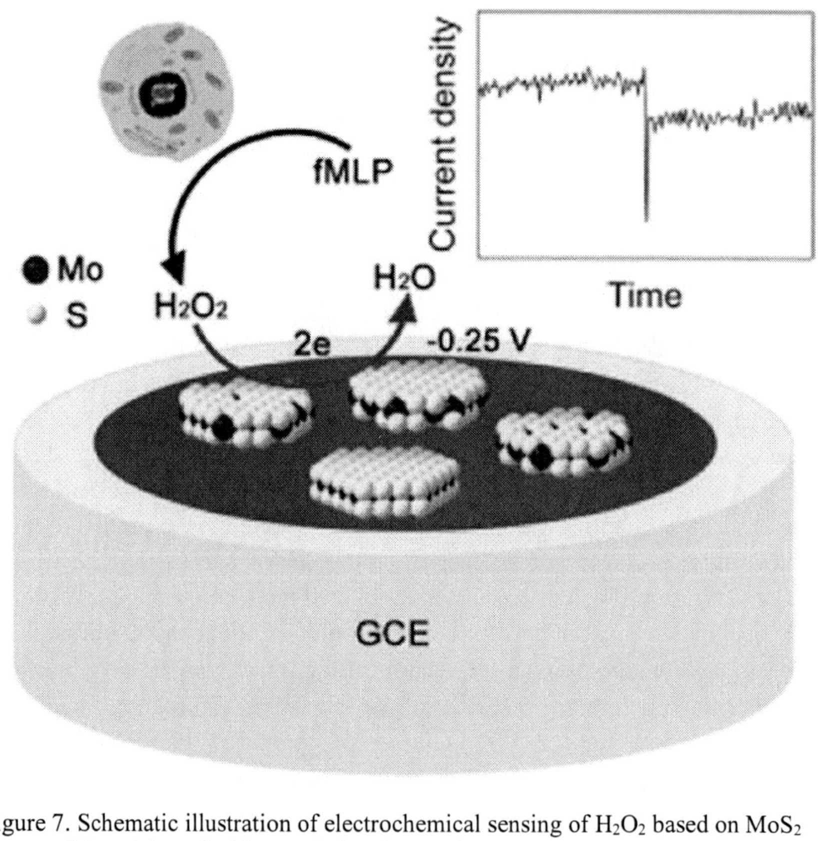

Figure 7. Schematic illustration of electrochemical sensing of H_2O_2 based on MoS_2 nanoparticles. Adapted with permission from ref. 83. Copyright 2013 American Chemical Society.

In the past few years, to further enhance the selectivity of the amperometric sensor based on TMD nanosheets, there are many investigations focused on DNA or aptamers modified on the electrode to detect various analytes of interest selectively. Huang et al. synthesized MoS_2/multi-walled carbon nanotube nanocomposite using hydrothermal method, and assembled probe DNA on the nanocomposite, glucose oxidase, and AuNP-modified GCE. Then, the assembled GCE was used to construct an amperometric sensor for detection of target DNA [109]. Wang et al. prepared MoS_2-thionin nanocomposite using ultrasonication and gradient centrifugation method, and exploited the nanocomposite to construct an amperometric sensor for detection of dsDNA on the basis of stable electrochemical response, intercalation, and electrostatic interaction of thionin with dsDNA (see Figure 8B) [110]. Yang's group combined the different affinity of MoS_2 nanosheets toward ssDNA versus dsDNA with methyl blue as electroactive indicator to develop an amperometric sensor for detection of target DNA [111]. Cao et al. assembled probe DNA on a GCE modified with a nanocomposite containing MoS_2, graphene, chitosan and AuNPs, and use thiol-tagged DNA coupled to HRP conjugated to AuNPs as a tracer to fabricate an amperometric sensor for detection of target DNA [112].

Huang's group reported the immobilization of the 17β-estradiol-binding aptamer onto the WS_2 nanosheets/Au nanoparticle-modified GCE, and the use of $[Fe(CN)_6]^{3-/4-}$ as the electrochemical indicator to construct an amperometric sensor for detection of 17β-estradiol (see Figure 8C) [113], and after that, they exploited the similar principle and used AuNPs as signal amplification for amperometric aptamer-based sensing for immunoglobulin E (IgE), platelet-derived growth factor BB (PDGF-BB) and target DNA [114-116]. In addition, Huang's group exploited the signal amplification of MoS_2/carbon aerogel nanocomposites and sandwich assay to develop a new amperometric aptamer-based sensor for detection of PDGF-BB [117]. To further improve detection sensitivity, Huang's group and Yuan's group exploited other nanocomposites based on TMD nanosheets and enzymatic signal amplification techniques to develop a new amperometric aptasensor for detection of PDGF-BB and DNA [118-121]. It was worth noting that for the first time, Loo et al had recently exploited MoS_2 nanoflakes as inherently electroactive labels, and combined the differential affinity of MoS_2 nanoflakes towards ssDNA and dsDNA to develop a new amperometric DNA hybridization sensor [122].

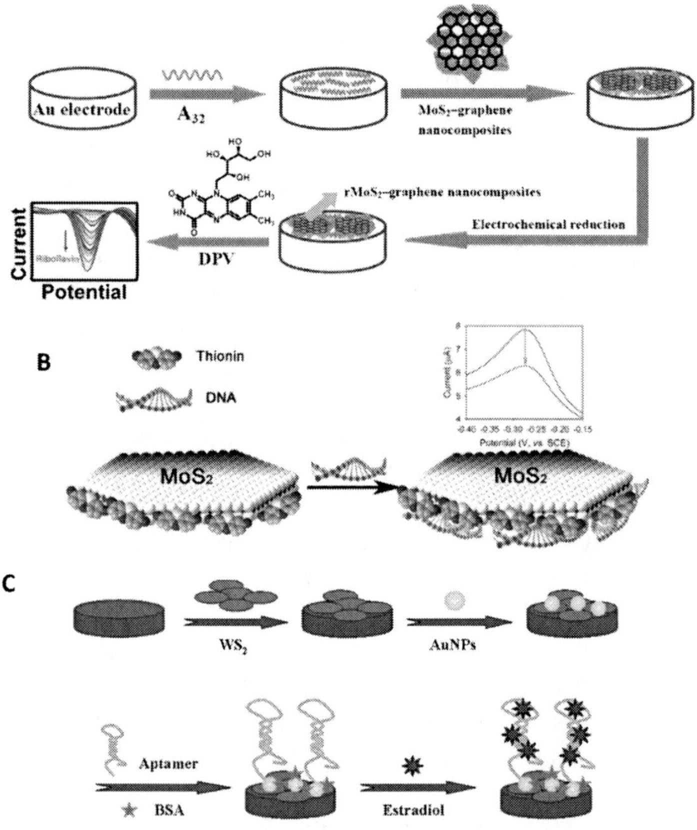

Figure 8. (A) Schematic illustration of fabrication process of the $rMoS_2$-graphene/A_{32}/Au electrode and its use for amperometric sensing of riboflavin. Adapted with permission from ref. 108. Copyright 2015 Elsevier. (B) Schematic diagram of the amperometric sensor for double-stranded DNA detection based on MoS_2-thionin nanocomposite. Adapted with permission from ref. 110. Copyright 2014 American Chemical Society. (C) Schematic diagram of the amperometric sensor for 17β-estradiol detection. Adapted with permission from ref. 113. Copyright 2014 Royal Society of Chemistry.

2.2.2. Impedimetric, Electrochemiluminescence and Photoelectrochemical Sensors

Compared with amperometric sensors, the impedimetric, electrochemiluminescence (ECL) and photoelectrochemical sensors based on TMD nanosheets are currently very rare. Pumera's group found that the impedance phase spectra of metallic 1T phase WS_2 exhibited specific resonant frequencies for water and methanol vapors. On the basis of the observation, they construct a sensor for detection of vapors based on impedance spectroscopy [123]. Liu's group presented a facile *in situ* synthesis route to prepare MoS_2-AuNPs nanocomposites, and exploited the nanocomposites as a matrix and quantum dots as a signal probe to construct a sandwich ECL aptasensor for detection of PDGF-BB [124], and after that, they synthesized hydrothermally flowerlike CdS-MoS_2 nanocomposites, immobilized the nanocomposites on GCE surface as ECL emitter, and then exploited IgE-binding aptamer and DNAzyme with HRP-like activity to construct a sandwich ECL sensor for IgE detection [125]. Yan's group applied MoS_2-AuNPs with excellent catalytic activity and CdS/ZnS quantum dots coated with polyethyleneimine with good ECL property to construct a simple sandwich-type ECL sensor for DNA detection [126]. Ou's group applied three-dimensional MoS_2-polyaniline nanoflowers, Ag nanocubes and cholesterol oxidase to fabricate a cathodic luminol-based ECL sensor for cholesterol detection [127]. Niu's group employed WS_2-titanium dioxide nanocomposites as optoelectronic materials in a novel photoelectrochemical platform to develop a sensor for accurate discrimination and detection of odiphenol and its derivatives [128]. Lei's group exploited the advantage of CdS/MoS_2 heterojunction, DNA recycling strategy and dual hemin-mediated chemiluminescence amplification to construct a photoelectrochemical sensor for detection of target DNA [129].

2.3. Field-Effect Transistor Sensors

Field-effect transistor (FET) sensor is one type of integrated device that changes its source-drain "channel" conductivity depending on the electric field of the environment. The FET sensor has recently captured huge attention due to its many desirable characteristics like portability, cheapness, low sample volume, short detection time, high sensitivity and high selectivity [130-153]. Very recently, TMD nanosheets have been found to have many advantages including large surface-to-volume ratio and good dielectric property [130-153]. The outstanding properties of TMD nanosheets facilitate the fabrication of FET sensors. Currently, many researchers have reported the use of FET sensors for detection of various analytes of interest [130-153].

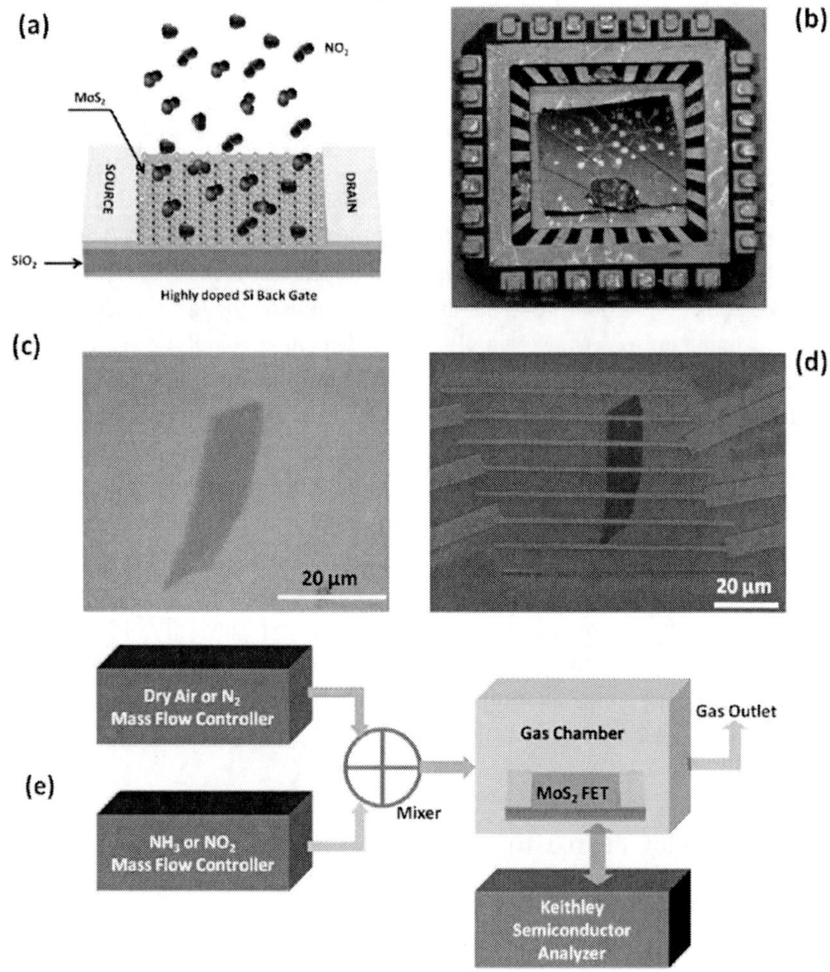

Figure 9. (a) Schematic illustration of MoS_2-based FET sensor for detection of NO_2 gas. (b) Optical image of the MoS_2-based sensing device mounted on the chip. (c) Optical image of two-layer MoS_2 nanosheet. (d) SEM image of two-layer MoS_2 transistor device. (e) Experimental setup for sensing of NH_3 and NO_2 gas. Adapted with permission from ref. 132. Copyright 2013 American Chemical Society.

Zhang's group exploited the MoS_2 film prepared by mechanical exfoliation method to construct a NO sensor [130]. Then, they developed a fully solution-processed fabrication of thin-film transistor array based on the MoS_2 thin film channel and reduced graphene oxide electrodes [131]. The obtained device was used to construct a FET sensor for NO detection. Functionalization of the MoS_2

thin film with Pt nanoparticles can further increase the sensitivity of the FET sensor. Late et al. constructed a FET sensor using the MoS_2 films with different thicknesses, and evaluated the transistors for gas sensing performances for NO_2, NH_3, and humidity with and without gate bias and light irradiation (see Figure 9) [132]. Wang and co-workers reported a sensor platform for detection of cancer marker proteins, prostate-specific antigens, using multilayer MoS_2 nanosheet-based field-effect device with versatile HfO_2-silane-based biofunctionalization scheme [133]. Sarkar et al. used MoS_2 nanosheets to construct a next generation label-free FET sensor for detection of pH and biomolecules. They showed that MoS_2-based FET sensor was superior to that based on graphene in terms of detection sensitivity because graphene had a lack of band gap [134]. Then, they investigated the doping effect of noble metals on TMD nanosheets for the first time, and found that the FET sensor based on MoS_2 nanosheets doped with Pd nanoparticles could be used for sensing of hydrogen gas [135]. Liu et al. demonstrated that CVD-grown MoS_2 nanosheets with Schottky contacts displayed current changes by 2-3 orders of magnitude upon exposure to NO_2 and NH_3 at a low concentration. On the basis of the observation, they constructed a FET sensor for detection of NO_2 and NH_3 based on MoS_2 nanosheets, and at the same time, they found that the improve sensitivity originated from the charge transfer between gas molecules and MoS_2 nanosheets and Schottky barrier modulation at MoS_2-metal electrode junctions [136]. Huo et al. demonstrated a WS_2 nanosheet-based FET sensor for detection of ethanol and NH_3 on the basis of the charge transfer between gas molecules and WS_2 nanosheets [137]. Benefiting from the charge transfer between gas molecules and TMD nanosheets, Late et al described the use of $MoSe_2$ nanosheets to construct a FET sensor for detection of NH_3 [138], and Xie et al. combined poly(3-hexylthiophene) and MoS_2 as sensitive materials to construct a FET sensor for detection of NH_3 [139]. Friedman and co-worker fabricated MoS_2-based FET sensor for detection of various analytes, and found that the sensor displayed the strongest response to triethylamine, a nerve gas by-product. The high sensitivity of the sensor for triethylamine was attributed to the electron-donor mechanism [140]. Lee et al. constructed a MoS_2-based FET sensor to detect prostate specific antigen. The sensor did not require a dielectric layer of insulating oxide owing to the hydrophobicity of MoS_2 [141]. Jin et al. used MoS_2 nanosheet-Pd nanoparticle composite to construct a FET sensor for H_2 detection, and found that compared with graphene-Pd composite, MoS_2-Pd composite exhibited much higher sensor response with shorter response and recovery times [142]. Samnakay et al. exploited MoS_2 thin-film to fabricate a FET sensor for detection of various vapors like ethanol, acetonitrile, toluene, chloroform, and methanol, and in the meantime, they found that transient time of the current change and the normalized spectral density of the low-frequency current fluctuations could be employed as additional sensing parameters for the FET sensor [143]. Nam et al. demonstrated that MoS_2-based transistor sensors were capable of probing the interaction between tumor necrosis factor-alpha (TNF-α) and its antibody, and were also used for detecting TNF-α [144]. Then, they exploited two different physics principles based on insulating-layer coated and insulating-layer-free MoS_2 FET for sensing of TNF-α [145]. Cho et al. reported the combination of MoS_2 and graphene to construct a FET sensor for detection of NO_2 gas molecules [146]. Yang's group prepared graphene/MoS_2 nanocomposite though an efficient co-exfoliation method, and used the nanocomposite to fabricate a FET sensor for methanol gas detection [147]. It was found the performance of the nanocomposite-based FET sensor surpassed that of the MoS_2-based sensor due to the synergetic effect provided by excellent electrocatalytic property of MoS_2 and superior carrier mobility of graphene. Lin et al. fabricated layered $MoTe_2$ transistors on Si/SiO_2 substrates, and found that the transistor showed ambipolar charge-transport behavior, resulting in its use for environmental sensors [148]. Zhu and co-worker developed a MoS_2-based FET sensor for O_2 detection, and studied the effect of gate voltage stress on the performance of the sensor. It was found that the use of gate voltage stress could improve the response and recovery of the O_2 sensor [149]. Donarelli's group fabricated MoS_2-based FET sensors on prepatterned Si_3N_4 substrates for detection of NO_2 and H_2 [150]. Jiang et al. reported a MoS_2-based FET sensing of Hg^{2+} [151]. The sensor is based on the p-type doping effect caused by the high binding affinity between Hg^{2+} and the sulfur sites on the surface of MoS_2 layers. Lee et al. demonstrated the use of ssDNA probe physically adsorbed on the MoS_2 FET for sensing of target DNA. The sensor was based on the electrostatic gating effect induced by the detachment of negatively charged probe DNA after hybridization from the MoS_2 channel surface [152]. Pawbake et al. synthesized WS_2 nanoparticles by using a modified hot wire chemical vapor deposition method, and exploited the nanomaterial to construct a humidity sensor [153].

2.4. Mass-Sensitive Sensors

In general, mass-sensitive sensors can be divided into two types: quartz crystal microbalance (QCM) sensors and piezoelectric sensors. At present, the development of mass-sensitive sensors has been still in the initial stage.

In 2012, Lieberzeit's group exploited MoS_2 nanoparticles as a recognition layer material to construct a QCM sensor for detection of thiol vapors [154]. It was found that the application of MoS_2 nanoparticles greatly improve the sensitivity of the sensor due to the soft metal sulfide-thiols affinity interactions. Bajwa et al. used MoS_2 nanoparticles for QCM-based sensing of 1-octanethiol vapors, and they also found that the use of MoS_2 nanoparticles could enhance the sensitivity of the sensor [155].

Duerloo et al. and Zhu et al. discovered that many TMD nanosheets possessed strong intrinsic piezoelectricity, which can be used to achieve electromechanical coupling in a rich variety of sensors [156, 157]. Park et al. reported a touch sensor with MoS_2/field effect transistors stacked with a piezoelectric polymer, polyvinylidene fluoride-trifluoroethylene [158]. Qi et al. gave the detailed investigation of the piezoelectric effect in monolayer MoS_2 devices under isotropic mechanical deformation, and found that the piezoelectric charge polarization-induced built-in electric field under strain variation could tune the conductivity of monolayer MoS_2 devices [159]. Depending on the intrinsic piezoelectricity of TMD nanosheets, Yu et al. designed a new piezoelectric sensors based on MoS_2 nanoribbon PN junctions and heterojunctions [160].

Conclusion

In the chapter, we summarized recent advances of TMD nanosheets in the sensing field. Benefiting from the unique and outstanding physicochemical properties of TMD nanosheets, many sensors with excellent performance were developed for detection of various analytes. Despite remarkable progress in the sensing field, there remain several challenges that limit their practical applications.

First of all, most of the reported sensors are currently employed under the optimal laboratory conditions as proof-of-concept, and they can be usually inappropriate for the applications in complex environment due to the presence of unknown background and interferents in real samples. To move this field forward, much work remains to be done. For example, to improve the sensing performance based on TMD nanosheets, it is of critical need to precisely control the synthesis of TMD nanosheets, and investigate the related synthesis mechanism in detail. Moreover, the selectivity of the sensor based on TMD nanosheets can be enhanced by introducing some functional materials such as aptamers, DNAzymes, peptides, antibodies and so on. More importantly, a good solution to the problem of the unknown background and interferents in real samples is to apply the trilinear three-way data (excitation-emission matrix data, kinetic-spectral data, pH-electrochemical data, and etc.) combined with second-order calibration methods (parallel factor analysis, multivariate curve resolution by alternating least-squares, alternating trilinear decomposition, and etc.). The second-order calibration approach can achieve the so-called second-order advantage, which in principle allows quantification of analytes in sample with unexpected components that are potential interferences.

Secondly, current sensors based on TMD nanosheets mainly concentrate on limited materials like MoS_2, WS_2, TaS_2. Therefore, some other TMD nanomaterials or the related hybrid nanomaterial with intriguing properties needs to be developed. In addition, the sensors based on TMD nanosheets mainly focus on the optical and electrochemical sensors, and other types of sensors like quartz crystal microbalance sensor, piezoelectric sensor, and thermal sensors are still in the initial stage. Therefore, there is an urgent need to develop other types of sensors.

Finally, most of the reported sensors based on TMD nanosheets currently focus on the detection of single analyte, and few literature reported the construction of multiplexed sensors for simultaneous detection of several analytes. Hence, it is expected that the development of the multiplexed sensors very likely become a hot topic in the field of sensing based on TMD nanosheets in the near future.

ACKNOWLEDGMENTS

This book chapter was supported by the National Natural Science Foundation of China (NSFC-21305061), the Natural Science Foundation of Jiangxi Province (20132BAB213011, 20151BAB213014 and 20151BAB203021), the Jiangxi Provincial Department of Education (GJJ13026), the State Key Laboratory of Chemo/Biosensing and Chemometrics of Hunan University (SKLCBC-2013010).

REFERENCES

[1] Geim, A. K. Graphene: status and prospects. *Science*, 2009, *vol.* 324(5934), 1530–1534.

[2] Castro Neto, A. H.; Guinea, F.; Peres, N. M. R.; Novoselov, K. S.; Geim, A. K. The electronic properties of graphene. *Rev. Mod. Phys.*, 2009, *vol.* 81(1), 109−162.

[3] Dreyer, D. R.; Park, S.; Bielawski, C. W.; Ruoff, R. S. The chemistry of graphene oxide. *Chem. Soc. Rev.*, 2010, 39(1), 228−240.

[4] Zhu, Y. W.; Murali, S.; Cai, W. W.; Li, X. S.; Suk, J. W.; Potts, J. R.; Ruoff, R. S. Graphene and graphene oxide: synthesis, properties, and applications. *Adv. Mater.*, 2010, *vol.* 22(35), 3906−3924.

[5] Bonaccorso, F.; Sun, Z.; Hasan, T.; Ferrari, A. C. Graphene photonics and optoelectronics. *Nat. Photonics*, 2010, *vol.* 4(9), 611−622.

[6] Allen, M. J.; Tung, V. C.; Kaner, R. B. Honeycomb carbon: a review of graphene. *Chem. Rev.*, 2010, *110*(1), vol. 132−145.

[7] Rao, C. N. R.; Sood, A. K.; Subrahmanyam, K. S.; Govindaraj, A. Graphene: the new two−dimensional nanomaterial. *Angew. Chem. Int. Ed.*, 2010, *vol.* 110(1), 132−145.

[8] Novoselov, K. S.; Fal'ko V. I.; Colombo, L.; Gellert, P. R.; Schwab, M. G.; Kim, K. A roadmap for graphene. *Nature*, 2012, *vol.* 490(7419), 192−200.

[9] Loh, K. P.; Bao, Q. L.; Eda, G.; Chhowalla, M. Graphene oxide as a chemically tunable platform for optical applications. *Nat. Chem.*, 2010, vol. 2(12), 1015−1024.

[10] Wu, J. S.; Pisula, W.; Mullen, K. Graphenes as potential material for electronics. *Chem. Rev.*, 2007, *vol.* 107(3), 718−747.

[11] Pumera, M.; Ambrosi, A.; Bonanni, A.; Chng, E. L. K.; Poh, H. L. Graphene for electrochemical sensing and biosensing. *Trends Anal. Chem.*, 2011, *vol.* 29(9), 954−965.

[12] Yang, K.; Feng, L. Z.; Liu, Z. The advancing uses of nano−graphene in drug delivery. *Expert Opin. Drug Discovery*, 2015, *vol.* 12(4), 601 612.

[13] Fan, X. L.; Chen, X. L.; Dai, L. M. 3D graphene based materials for energy storage. *Curr. Opin. Colloid Interface Sci.*, 2015, *vol.* 20(5−6), 429−438.

[14] Cao, L.; Meziani, M. J.; Sahu, S.; Sun, Y. P. Photoluminescence properties of graphene versus other carbon nanomaterials. *Acc. Chem. Res.*, 2013, *vol.* 46(1), 171−180.

[15] Tzalenchuk, A.; Lara−Avila, S.; Kalaboukhov, A.; Paolillo, S.; Syvajarvi, M.; Yakimova, R.; Kazakova, O.; Janssen, T. J. B. M.; Fal'ko, V.; Kubatkin, S. Boron nitride substrates for high−quality graphene electronics. *Nat. Nanotechnol.*, 2010, *vol.* 5(10), 186−189.

[16] Bhimanapati, G. R.; Lin, Z.; Meunier, V.; Jung, Y.; Cha, J.; Das, S.; Xiao, D.; Son, Y.; Strano, M. S.; Cooper, V. R.; Liang, L. B.; Louie, S. G.; Ringe, E.; Zhou, W.; Kim, S. S.; Naik, R. R.; Sumpter, B. G.; Terrones, H.; Xia, F. N.; Wang, Y. L.; Zhu, J.; Akinwande, D.; Alem, N.; Schuller, J. A.; Schaak, R. E.; Terrones, M; Robinson, J. A. Recent advances in two–dimensional materials beyond graphene. *ACS Nano*, 2015, *vol.* 9(12), 11509–11539.

[17] Jariwala, D.; Sangwan, V. K.; Lauhon, L. J.; Marks, T. J.; Hersam, M. C. Emerging device applications for semiconducting two–dimensional transition metal dichalcogenides. *ACS Nano*, 2013, *vol.* 5(4), 263–275.

[18] Chhowalla, M.; Shin, H. S.; Eda, G.; Li, L. J.; Loh, K. P.; Zhang, H. The chemistry of two–dimensional layered transition metal dichalcogenide nanosheets. *Nat. Chem.*, 2013, *vol.* 5(4), 263–275.

[19] Rao, C. N. R.; Gopalakrishnan, K.; Maitra, U. Comparative study of potential applications of graphene, MoS_2, and other two–dimensional materials in energy devices, sensors, and related areas. *ACS Appl. Mater. Interfaces*, 2015, *vol.* 7(15), 7809–7832.

[20] Tan, C. L.; Zhang, H. Two–dimensional transition metal dichalcogenide nanosheet–based composites. *Chem. Soc. Rev.*, 2015, vol. 44(9), 2713–2731.

[21] Lembke, D.; Bertolazzi, S.; Kis, A. Single–layer MoS_2 electronics. *Acc. Chem. Res.*, 2015, *vol.* 48(1), 100–110.

[22] Zhang, X.; Qiao, X. F.; Shi, W.; Wu, J. B.; Jiang, D. S.; Tan, P. H. Phonon and Raman scattering of two–dimensional transition metal dichalcogenides from monolayer, multilayer to bulk material. *Chem. Soc. Rev.*, 2015, *vol.* 44(9), 2757–2785.

[23] Wang, F. M.; Shifa, T. A.; Zhan, X. Y.; Huang, Y.; Liu, K. L.; Cheng, Z. Z.; Jiang, C.; He, J. Recent advances in transition–metal dichalcogenide based nanomaterials for water splitting. *Nanoscale*, 2015, *vol.* 7(47), 19764–19788.

[24] Li, H. N.; Shi, Y. M.; Chiu, M. H.; Li, L. J. Emerging energy applications of two–dimensional layered transition metal dichalcogenides. *Nano Energy*, 2015, *vol.* 18, 293–305.

[25] Wang, H.; Feng, H. B.; Li, J. H. Graphene and graphene–like layered transition metal dichalcogenides in energy conversion and storage. *Small*, 2014, *vol.* 10(11), 2165–2181.

[26] Yang, J.; Shin, H. S. Recent advances in layered transition metal dichalcogenides for hydrogen evolution reaction. *J. Mater. Chem. A*, 2014, *vol.* 2(17), 5979–5985.

[27] Chia, X. Y.; Ambrosi, A.; Sofer, Z.; Luxa, J.; Pumera, M. Catalytic and charge transfer properties of transition metal dichalcogenides arising from electrochemical pretreatment. *ACS Nano*, 2015, *vol.* 9(5), 5164–5179.

[28] Radisavljevic, B.; Radenovic, A.; Brivio, J.; Giacometti, V.; Kis, A. Single–layer MoS_2 transistors. *Nat. Nanotechnol.*, 2011, *vol.* 6(3), 147–150.

[29] Pumera, M.; Loo, A. H. Layered transition–metal dichalcogenides (MoS_2 and WS_2) for sensing and biosensing. *Trends Anal. Chem.*, 2014, *vol.* 61, 49–53.

[30] Huang, Y. X.; Guo, J. H.; Kang, Y. J.; Ai, Y.; Li, C. M. Two dimensional atomically thin MoS_2 nanosheets and their sensing applications. *Nanoscale*, 2015, *vol.* 7(46), 19358–19376.

[31] Zhang, W. S.; Zhang, P. P.; Su, Z. Q.; Wei, G. Synthesis and sensor applications of MoS_2–based nanocomposites. *Nanoscale*, 2015, *vol.* 7(44), 18364–18378.

[32] Janata, J. *Principles of Chemical Sensors*, 2nd ed.; Springer: New York, 2009.

[33] Grundler, P. *Chemical Sensors: An Introduction for Scientists and Engineers*, 1st ed.; Springer: New York, 2007.

[34] Faaden, J. *Handbook of Modern Sensors: Physics, Designs, and Applications*, 1st ed.; Springer: New York, 2010.

[35] Gao, L. Z.; Zhuang, J.; Nie, L.; Zhang, J. B.; Zhang, Y.; Gu, N.; Wang, T. H.; Feng, J.; Yang, D. L.; Perrett, S.; Yan, X. Y. Intrinsic peroxidase–like activity of ferromagnetic nanoparticles. *Nat. Nanotechnol.*, 2007, *vol.* 2(9), 577–583.

[36] Wei, H.; Wang, E. K. Nanomaterials with enzyme–like characteristics (nanozymes): next–generation artificial enzymes. *Chem. Soc. Rev.*, 2013, *vol.* 42(14), 6060–6093.

[37] Garg, B.; Bisht, T.; Ling, Y. C. Graphene–based nanomaterials as efficient peroxidase mimetic catalysts for biosensing applications: an overview. *Molecules*, 2015, *vol.* 20(8), 14155–14190.

[38] Wang, X. Y.; Hu, Y. H.; Wei, H. Nanozymes in bionanotechnology: from sensing to therapeutics and beyond. *Inorg. Chem. Front.*, 2016, *vol.* 3(1), 41–60.

[39] Lin, T. R.; Zhong, L. S.; Guo, L. Q.; Fu, F. F.; Chen, G. N. Seeing diabetes: visual detection of glucose based on the intrinsic peroxidase–like activity of MoS_2 nanosheets. *Nanoscale*, 2014, *vol.* 6(20), 11856−11862.

[40] Lin, T. R.; Zhong, L. S.; Song, Z. P.; Guo, L. Q.; Wu, H. Y.; Guo, Q. Q.; Chen, Y.; Fu, F. F.; Chen, G. N. Visual detection of blood glucose based on peroxidase–like activity of WS_2 nanosheets. *Biosens. Bioelectron.*, 2014, *vol.* 62, 302−307.

[41] Chen, Q.; Chen, J.; Gao, C. J.; Zhang, M. L.; Chen, J. Y.; Qiu, H. D. Hemin–functionalized WS_2 nanosheets as highly active peroxidase mimetics for label–free colorimetric detection of H_2O_2 and glucose. *Analyst*, 2015, *vol.* 140(8), 2857−2863.

[42] Guo, X. R.; Wang, Y.; Wu, F. Y.; Ni, Y. N.; Kokot, S. A colorimetric method of analysis for trace amounts of hydrogen peroxide with the use of the nano–properties of molybdenum disulfide. *Analyst*, 2015, *vol.* 140(4), 1119−1126.

[43] Zhao, K.; Gu, W.; Zheng, S. S.; Zhang, C. L.; Xian, Y. Z. SDS–MoS_2 nanoparticles as highly–efficient peroxidase mimetics for colorimetric detection of H_2O_2 and glucose. *Talanta*, 2015, *vol.* 141, 47−52.

[44] Lei, J. Y.; Lu, X. F.; Nie, G. D.; Jiang, Z. Q.; Wang, C. One–pot synthesis of algae–like MoS_2/PPy nanocomposite: a synergistic catalyst with superior peroxidase–like catalytic activity for H_2O_2 detection. *Part. Part. Syst. Char.*, 2015, *vol.* 32(9), 886−892.

[45] Nirala, N. R.; Pandey, S.; Bansal, A.; Singh, V. K.; Mukherjee, B.; Saxena, P. S.; Srivastava, A. Different shades of cholesterol: Gold nanoparticles supported on MoS_2 nanoribbons for enhanced colorimetric sensing of free cholesterol. *Biosens. Bioelectron.*, 2015, *vol.* 74, 207−213.

[46] Cai, S. F.; Han, Q. S.; Qi, C.; Lian, Z.; Jia, X. H.; Yang, R.; Wang, C. $Pt_{74}Ag_{26}$ nanoparticle–decorated ultrathin MoS_2 nanosheets as novel peroxidase mimics for highly selective colorimetric detection of H_2O_2 and glucose. *Nanoscale*, 2016, *vol.* 8(6), 3685−3693.

[47] Wang, Y.; Hu, J.; Zhuang, Q. F.; Ni, Y. N. Enhancing sensitivity and selectivity in a label–free colorimetric sensor for detection of iron(II) ions with luminescent molybdenum disulfide nanosheet–based peroxidase mimetics. *Biosens. Bioelectron.*, 2016, *vol.* 80, 111−117.

[48] Lakowicz, J. R. *Principles of Fluorescence Spectroscopy*, 3rd ed.; Springer: New York, 2006.

[49] Resch–Genger, U.; Grabolle, M.; Cavaliere–Jaricot, S.; Nitschke, R.; Nann, T. Quantum dots versus organic dyes as fluorescent labels. *Nat. Methods*, 2008, *vol.* 5(9), 763–775.

[50] Massey, M.; Algar, W. R.; Krull, U. J. Fluorescence resonance energy transfer (FRET) for DNA biosensors: FRET pairs and Förster distances for various dye–DNA conjugates. *Anal. Chim.* Acta, 2006, *vol.* 568(1–2), 181–189.

[51] Shi, J. Y.; Tian, F.; Lyu, J.; Yang, M. Nanoparticle based fluorescence resonance energy transfer (FRET) for biosensing applications. *J. Mater. Chem. B*, 2015, *vol.* 3(35), 6989–7005.

[52] Stanisavljevic, M.; Krizkova, S.; Vaculovicova, M.; Kizek, R.; Adam, V. Quantum dots–fluorescence resonance energy transfer–based nanosensors and their application. *Biosens. Bioelecton*, 2015, *vol.* 74, 562–574.

[53] Zhong, W. W. Nanomaterials in fluorescence–based biosensing. *Anal. Bioanal. Chem.*, 2009, *vol.* 394(1), 47–59.

[54] Jennings, T. L.; Singh, M. P.; Strouse, G. F. Fluorescent lifetime quenching near *d*=1.5 nm gold nanoparticles: probing NSET validity. *J. Am. Chem. Soc.*, 2006, *vol.* 128(16), 5462–5467.

[55] Li, M.; Cushing, S. K.; Wang, Q. Y.; Shi, X. D.; Hornak, L. A.; Hong, Z. L.; Wu, N. Q. Size–dependent energy transfer between CdSe/ZnS quantum dots and gold nanoparticles. *J. Phys. Chem. Lett.*, 2011, *vol.* 2(17), 2125–2129.

[56] Li, C.; Shi, G. Q. Carbon nanotube–based fluorescence sensors. *J. Photochem. Photobiol. C*, 2014, *vol.* 19(16), 20–34.

[57] Swathi, R. S.; Sebastian, K. L. Long range resonance energy transfer from a dye molecule to graphene has (distance)$^{-4}$ dependence. *J. Chem. Phys.*, 2009, *vol.* 130(8), 086101.

[58] Swathi, R. S.; Sebastian, K. L. Resonance energy transfer from a dye molecule to graphene. *J. Chem. Phys.*, 2008, *vol.* 129(5), 054703.

[59] Zhu, C. F.; Zeng, Z. Y.; Li, H.; Li, F.; Fan, C. H.; Zhang, H. Single–layer MoS_2–based nanoprobes for homogeneous detection of biomolecules. *J. Am. Chem. Soc.*, 2013, *vol.* 135(16), 5998–6001.

[60] Ge, J.; Qu, E. C.; Yu, R. Q.; Chu, X. A novel aptameric nanobiosensor based on the self–assembled DNA–MoS_2 nanosheet architecture for biomolecule detection. *J. Mater. Chem. B*, 2014, *vol.* 2(6), 625–628.

[61] Jia, L.; Ding, L.; Tian, J. W.; Bao, L.; Hu, Y. P.; Ju, H. X.; Yu, J. S. Aptamer loaded MoS_2 nanoplates as nanoprobes for detection of intracellular ATP and controllable photodynamic therapy. *Nanoscale*, 2015, *vol.* 7(38), 15953−15961.

[62] Kong, R. M.; Ding, L.; Wang, Z. J.; You, J. M.; Qu, F. L. A novel aptamer–functionalized MoS_2 nanosheet fluorescent biosensor for sensitive detection of prostate specific antigen. *Anal. Bioanal. Chem.*, 2015, *vol.* 407(2), 369−377.

[63] Mao, K.; Wu, Z. T.; Chen, Y. R.; Zhou, X. D.; Shen, A. G.; Hu, J. M. A novel biosensor based on single–layer MoS_2 nanosheets for detection of Ag^+. *Talanta*, 2015, *vol.* 132, 658−663.

[64] Zhang, H. Y.; Ruan, Y. J.; Lin, L.; Lin, M. G.; Zeng, X. X.; Xi, Z. M.; Fu, F. F. A turn–off fluorescent biosensor for the rapid and sensitive detection of uranyl ion based on molybdenum disulfide nanosheets and specific DNAzyme. *Spectrochim. Acta, Part A*, 2015, *vol.* 146(6), 1−6.

[65] Huang, Y. X.; Shi, Y. M.; Yang, H. Y.; Ai, Y. A novel single–layered MoS_2 nanosheet based microfluidic biosensor for ultrasensitive detection of DNA. *Nanoscale*, 2015, *vol.* 7(6), 2245−2249.

[66] Yuan, Y. X.; Li, R. Q.; Liu, Z. H. Establishing water–soluble layered WS_2 nanosheet as a platform for biosensing. *Anal. Chem.*, 2014, *vol.* 86(7), 3610−3615.

[67] Yin, X. H.; Cai, J.; Feng, H. Y.; Wu, Z. M.; Zou, J. M.; Cai, Q. Y. A novel VS_2 nanosheet–based biosensor for rapid fluorescence detection of cytochrome c. New *J. Chem.*, 2015, *vol.* 39(3), 1892−1898.

[68] Tan, C. L.; Yu, P.; Hu, Y. L.; Chen, J. Z.; Huang, Y.; Cai, Y. Q.; Luo, Z. M.; Li, B.; Lu, Q. P.; Wang, L. H.; Liu, Z.; Zhang, H. High–yield exfoliation of ultrathin two–dimensional ternary chalcogenide nanosheets for highly sensitive and selective fluorescence DNA sensors. *J. Am. Chem. Soc.*, 2015, *vol.* 137(32), 10430−10436.

[69] Zhang, Y.; Zheng, B.; Zhu, C. F.; Zhang, X.; Tan, C. L.; Li, H.; Chen, B.; Yang, J.; Chen, J. Z.; Huang, Y.; Wang, L. H.; Zhang, H. Single–layer transition metal dichalcogenide nanosheet–based nanosensors for rapid, sensitive, and multiplexed detection of DNA. *Adv. Mater.*, 2015, *vol.* 27(5), 935−939.

[70] Wang, S. T.; Zhang, Y. L.; Ning, Y.; Zhang, G. J. A WS_2 nanosheet–based platform for fluorescent DNA detection via PNA–DNA hybridization. *Analyst*, 2015, *vol.* 140(2), 434−439.

[71] Ge, J. Tang, L. J.; Xi, Q.; Li, X. P.; Yu, R. Q.; Jiang, J. H.; Chu, X. A WS_2 nanosheet based sensing platform for highly sensitive detection of T4 polynucleotide kinase and its inhibitors. *Nanoscale*, 2014, *vol.* 6(12), 6866–6872.

[72] Huang, J. H.; Ye, L.; Gao, X.; Li, H.; Xu, J. B.; Li, Z. G. Molybdenum disulfide–based amplified fluorescence DNA detection using hybridization chain reactions. *J. Mater. Chem. B*, 2015, *vol.* 3(11), 2395–2401.

[73] Deng, H. M.; Yang, X. J. Gao, Z. Q. MoS_2 nanosheets as an effective fluorescence quencher for DNA methyltransferase activity detection. *Analyst*, 2015, *vol.* 140(9), 3210–3215.

[74] Xi, Q., Zhou, D. M.; Kan, Y. Y.; Ge, J.; Wu, Z. K.; Yu, R. Q.; Jiang, J. H. Highly sensitive and selective strategy for MicroRNA detection based on WS_2 nanosheet mediated fluorescence quenching and duplex–specific nuclease signal amplification. *Anal. Chem.*, 2014, *vol.* 86(3), 1361–1365.

[75] Qin, Y. F.; Ma, Y. F.; Jin, X.; Zhang, L. L.; Ye, G. J.; Zhao, S. L. A sensitive fluorescence turn–on assay of bleomycin and nuclease using WS_2 nanosheet as an effective sensing platform. *Anal. Chim. Acta*, 2015, *vol.* 866, 84–89.

[76] Wang, Y.; Ni, Y. N. Molybdenum disulfide quantum dots as a photoluminescence sensing platform for 2,4,6–trinitrophenol detection. *Anal. Chem*, 2014, *vol.* 86(15), 7463–7470.

[77] Yang, Y.; Liu, T.; Cheng, L.; Song, G. S.; Liu, Z.; Chen, M. W. MoS_2–based nanoprobes for detection of silver ions in aqueous solutions and bacteria. *ACS Appl. Mater. Interfaces*, 2015, *vol.* 7(14), 7526–7533.

[78] Loan, P. T. K.; Zhang, W. J.; Lin, C. T.; Wei, K. H.; Li, L. J.; Chen, C. H. Graphene/MoS_2 heterostructures for ultrasensitive detection of DNA hybridisation. *Adv. Mater.*, 2014, *vol.* 26(28), 4838–4844.

[79] Wang, Y.; Hu, J.; Zhuang, Q. F.; Ni, Y. N. Label–free fluorescence sensing of lead(II) ions and sulfide ions based on luminescent molybdenum disulfide nanosheets. *ACS Sustainable Chem. Eng.*, 2016, *vol.* 4(5), 2535–2541.

[80] Hu, J.; Zhuang, Q. F.; Wang, Y.; Ni, Y. N. Label–free fluorescent catalytic biosensor for highly sensitive and selective detection of the ferrous ion in water samples using a layered molybdenum disulfide nanozyme coupled with an advanced chemometric model. *Analyst*, 2016, *vol.* 141(5), 1822–1829.

[81] Zhao, J. J.; Jin, X.; Vdovenko, M.; Zhang, L. L.; Sakharov, I. Y.; Zhao, S. L. A WS_2 nanosheet based chemiluminescence resonance energy transfer platform for sensing biomolecules. *Chem. Commun.*, 2015, *vol.* 51(55), 11092−11095.

[82] Zeng, S. W.; Hu, S. Y.; Xia, J.; Anderson, T.; Dinh, X. Q.; Meng, X. M.; Coquet, P.; Yong, K. T. Graphene–MoS_2 hybrid nanostructures enhanced surface plasmon resonance biosensors. *Sens. Actuators*, B, 2015, *vol.* 207, 801−810.

[83] Wang, T. Y.; Zhu, H. C.; Zhuo, J. Q.; Zhu, Z. W.; Papakonstantinou, P.; Lubarsky, G.; Lin, J.; Li, M. X. Biosensor based on ultrasmall MoS_2 nanoparticles for electrochemical detection of H_2O_2 released by cells at the nanomolar level. *Anal. Chem.*, 2013, *vol.* 85(21), 10289−10295.

[84] Huang, K. J.; Wang, L.; Li, J.; Liu, Y. M. Electrochemical sensing based on layered MoS_2−graphene composites. *Sens. Actuators*, *B*, 2013, *vol.* 178, 671−677.

[85] Huang, K. J.; Wang, L.; Li, Y. J.; Gan, T.; Liu, Y. M.; Wang, L. L.; Fan, Y. Synthesis and electrochemical performances of layered tungstensulfide–graphene nanocomposite as a sensing platform for catechol,resorcinol and hydroquinone. *Electrochim. Acta*, 2013, *vol.* 107, 379−387.

[86] Huang, J. W.; Dong, Z. P.; Li, Y. R.; Li, J.; Tang, W. J.; Yang, H. D.; Wang, J.; Bao, Y.; Jin, J.; Li, R. MoS_2 nanosheet functionalized with Cu nanoparticles and its application for glucose detection. *Mater. Res. Bull.*, 2013, *vol.* 48(11), 4544−4547.

[87] Su, S.; Sun, H. F.; Xu, F.; Yuwen, L. H.; Fan, C. H.; Wang, L. H. Direct electrochemistry of glucose oxidase and a biosensor for glucose based on a glass carbon electrode modified with MoS_2 nanosheets decorated with gold nanoparticles. *Microchim. Acta*, 2014, *vol.* 181(13), 1497−1503.

[88] Huang, K. J.; Liu, Y. J.; Liu, Y. M.; Wang, L. L. Molybdenum disulfide nanoflower–chitosan–Au nanoparticlescomposites based electrochemical sensing platform for bisphenol A determination. *J. Hazard. Mater.*, 2014, *vol.* 276, 207−215.

[89] Huang, K. J.; Zhang, J. Z.; Liu, Y. J.; Wang, L. L. Novel electrochemical sensing platform based on molybdenum disulfide nanosheets–polyaniline composites and Au nanoparticles. *Sens. Actuators*, *B*, 2014, *vol.* 194, 303−310.

[90] Sun, H. F.; Chao, J.; Zuo, X. L.; Su, S.; Liu, X. F.; Yuwen L. H.; Fan, C. H.; Wang, L. H. Gold nanoparticle–decorated MoS_2 nanosheets for simultaneous detection of ascorbic acid, dopamine and uric acid. *RSC Adv.*, 2014, *vol.* 4(52), 27625−27629.

[91] Huang, J. W.; He, Y. Q.; Jin, J.; Li, Y. R.; Dong, Z. P.; Li, R. A novel glucose sensor based on MoS_2 nanosheet functionalized with Ni nanoparticles. *Electrochim. Acta*, 2014, *vol.* 136, 41−46.

[92] Zhao, X. J.; Xia, X. H.; Yu, S. J.; Wang, C. M. An electrochemical sensor for honokiol based on a glassy carbon electrode modified with MoS_2/graphene nanohybrid film. *Anal. Methods*, 2014, *vol.* 6(23), 9375−9382.

[93] Feng, Q. L.; Duan, K. Y.; Ye, X. L.; Lu, D. B.; Du, Y. L.; Wang, C. M. A novel way for detection of eugenol via poly(diallyldimethylammonium chloride) functionalized graphene–MoS_2 nano–flower fabricated electrochemical sensor. *Sens. Actuators, B*, 2014, *vol.* 192, 1−8.

[94] Yang, T.; Chen, H. Y.; Yang, R. R.; Jiang, Y. H.; Li, W. H.; Jiao, K. A glassy carbon electrode modified with a nanocomposite consisting of molybdenum disulfide intercalated into self–doped polyaniline for the detection of bisphenol A. *Microchim. Acta*, 2015, *vol.* 182(15), 2623−2628.

[95] Zheng, Z. X.; Feng, Q. L.; Li, J.; Wang, C. M. Highly sensitive determination of chloramphenicol based on thin–layered MoS_2/polyaniline nanocomposite. *Talanta*, 2015, *vol.* 144, 1324−1328.

[96] Yang, T.; Chen, H. Y.; Ge, T.; Wang, J.; Li, W. H.; Jiao, K. The p–type MoS_2 nanocube modified poly(diallyl dimethyl ammonium chloride)–mesoporous carbon composites as a catalytic amplification platform for electrochemical detection of L–cysteine. *Sens. Actuators, B*, 2015, *vol.* 221, 1162−1169.

[97] Lin, Y. P.; Chen, X.; Lin, Y. X.; Zhou, Q.; Tang, D. P. Non–enzymatic sensing of hydrogen peroxide using a glassy carbon electrode modified with a nanocomposite made from carbon nanotubes and molybdenum disulfide. *Microchim. Acta*, 2015, *vol.* 182(9–10), 1803−1809.

[98] Kim, H. U.; Kim, H.; Ahn, C.; Kulkarni, A.; Jeon, M.; Yeom, G. Y.; Lee, M. H.; Kim, T. In situ synthesis of MoS_2 on a polymer based gold electrode platform and its application in electrochemical biosensing. *RSC Adv*, 2015, *vol.* 5(14), 10134−10138.

[99] Chao, J.; Zhou, M.; Zhang, C.; Sun, H. F.; Pan, D.; Pei, H.; Su, S.; Yuwen, L. H.; Fan, C. H.; Wang, L. H. A MoS_2–based system for efficient immobilization of hemoglobin and biosensing applications. *Nanotechnology*, 2015, *vol.* 26(27), 274005.
[100] Pramoda, K.; Moses, K.; Maitra, U.; Rao, C. N. R. Superior performance of a MoS_2–RGO composite and a borocarbonitride in the electrochemical detection of dopamine, uric Acid and adenine. *Electroanalysis*, 2015, *vol.* 27(8), 1892−1898.
[101] Yang, T.; Yang, R. R.; Chen, H. Y.; Nan, F. X.; Ge, T.; Jiao, K. Electrocatalytic activity of molybdenum disulfide nanosheets enhanced by self–doped polyaniline for highly sensitive and synergistic determination of adenine and guanine. *ACS Appl. Mater. Interfaces*, 2015, *vol.* 7(4), 2867−2872.
[102] Gan, X. R.; Zhao, H. M.; Chen, S.; Yu, H. T.; Quan, X. Three–dimensional porous H_xTiS_2 nanosheet−polyaniline nanocomposite electrodes for directly detecting trace Cu(II) ions. *Anal. Chem.*, 2015, *vol.* 87(11), 5605−5613.
[103] Chao, J.; Han, X. Y.; Sun, H. F.; Su, S.; Weng, L. X.; Wang, L. H. Platinum nanoparticles supported MoS_2 nanosheet for simultaneous detection of dopamine and uric acid. *Sci. China Chem.*, 2016, *vol.* 59(3), 332−337.
[104] Zhang, Y.; Chen, P.; Wen, F. F.; Yuan, B.; Wang, H. G. Fe_3O_4 nanospheres on MoS_2 nanoflake: electrocatalysis and detection of Cr(VI) and nitrite. *J. Electroanal. Chem.*, 2016, *vol.* 761, 14−20.
[105] Vasilescu, I.; Eremia, S. A. V.; Kusko, M.; Radoi, A.; Vasile, E.; Radu, G. L. Molybdenum disulphide and graphene quantum dots as electrode modifiers for laccase biosensor. *Biosens. Bioelectron.*, 2016, *vol.* 75, 232−237.
[106] Zhu, L. L.; Zhang, Y.; Xu, P. C.; Wen, W. J.; Li, X. X.; Xu, J. Q. PtW/MoS_2 hybrid nanocomposite for electrochemical sensingof H_2O_2 released from living cells. *Biosens. Bioelectron.*, 2016, *vol.* 80, 601−606.
[107] Wu, S. X.; Zeng, Z. Y.; He, Q. Y.; Wang, Z. J.; Wang, S. J.; Du, Y. P.; Yin, Z. Y.; Sun, X. P.; Chen, W.; Zhang, H. Electrochemically reduced single–layer MoS_2 nanosheets: characterization, properties, and sensing applications. *Small*, 2012, *vol.* 8(14), 2264−2270.

[108] Wang, Y.; Zhuang, Q. F.; Ni, Y. N. Fabrication of riboflavin electrochemical sensor based on homoadenine single–stranded DNA/molybdenum disulfide–graphene nanocomposite modified gold electrode. *J. Electroanal. Chem.*, 2015, *vol.* 736, 47–54.

[109] Huang, K. J.; Liu, Y. J.; Wang, H. B.; Wang, Y. Y.; Liu, Y. M. Sub–femtomolar DNA detection based on layered molybdenum disulfide/multi–walled carbon nanotube composites, Au nanoparticle and enzyme multiple signal amplification. *Biosens. Bioelectron*, 2014, *vol.* 55, 195–202.

[110] Wang, T. Y.; Zhu, R. Z.; Zhuo, J. Q.; Zhu, Z. W.; Shao, Y. H.; Li, M. X. Direct detection of DNA below ppb level based on thionin–functionalized layered MoS_2 electrochemical sensors. *Anal. Chem.*, 2014, *vol.* 86(24), 12064–12069.

[111] Wang, X. X.; Nan, F. X.; Zhao, J. L.; Yang, T.; Ge, T.; Jiao, K. A label–free ultrasensitive electrochemical DNA sensor based on thin–layer MoS_2 nanosheets with high electrochemical activity. *Biosens. Bioelectron.*, 2015, *vol.* 64, 386–391.

[112] Cao, X. Y. Ultra–sensitive electrochemical DNA biosensor based on signal amplification using gold nanoparticles modified with molybdenum disulfide, graphene and horseradish peroxidase. *Microchim. Acta*, 2014, *vol.* 181(9–10), 1133–1141.

[113] Huang, K. J.; Liu, Y. J.; Zhang, J. Z.; Liu, Y. M. A novel aptamer sensor based on layered tungsten disulfide nanosheets and Au nanoparticles amplification for 17β–estradiol detection. *Anal. Methods*, 2014, *vol.* 6(19), 8011–8017.

[114] Huang, K. J.; Liu, Y. J.; Cao, J. T.; Wang, H. B. An aptamer electrochemical assay for sensitive detection of immunoglobulin E based on tungsten disulfide–graphene composites and gold nanoparticles. *RSC Adv.*, 2014, *vol.* 4(69), 36742–36748.

[115] Liu, X.; Shuai, H. L.; Huang, K. J. A label–free electrochemical aptasensor based on leaf–like vanadium disulfide–Au nanoparticles for the sensitive and selective detection of platelet–derived growth factor BB. *Anal. Methods*, 2015, *vol.* 7(19), 8277–8284.

[116] Fang, L. X.; Cao, J. T.; Huang, K. J. A sensitive electrochemical biosensor for specific DNA sequence detection based on flower–like VS_2, graphene and Au nanoparticles signal amplification. *J. Electroanal. Chem.*, 2015, *vol.* 746, 1–8.

[117] Fang, L. X.; Huang, K. J.; Liu, Y. Nove lelectrochemical dual–aptamer–based sandwich biosensor using molybdenum disulfide/carbon aerogel composites and Au nano–particles for signal amplification. *Biosens. Bioelectron.*, 2015, *vol.* 71, 171–178.

[118] Huang, K. J.; Liu, Y. J.; Zhai, Q. F. Ultrasensitive biosensing platform based on layered vanadium disulfide–graphene composites coupling with tetrahedron–structured DNA probes and exonuclease III assisted signal amplification. *J. Mater. Chem. B*, 2015, *vol.* 3(41), 8180–8187.

[119] Huang, K. J.; Shuai, H. L.; Zhang, J. Z. Ultrasensitive sensing platform for platelet–derived growth factor BB detection based on layered molybdenum selenide–graphene composites and Exonuclease III assisted signal amplification. *Biosens. Bioelectron.*, 2016, *vol.* 77, 69–75.

[120] Shuai, H. L.; Huang, K. J.; Chen, Y. X. A layered tungsten disulfide/acetylene black composite based DNA biosensing platform coupled with hybridization chain reaction for signal amplification. *J. Mater. Chem. B*, 2016, *vol.* 4(6), 1186–1196.

[121] Jing, P.; Yi, H. Y.; Xue, S. Y.; Chai, Y. Q.; Yuan, R.; Xu, W. J. A sensitive electrochemical aptasensor based on palladium nanoparticles decorated graphene–molybdenum disulfide flower–like nanocomposites and enzymatic signal amplification. *Anal. Chim. Acta*, 2015, *vol.* 853(6), 234–241.

[122] Loo, A. H.; Bonanni, A.; Ambrosi, A.; Pumera, M. Molybdenum disulfide (MoS_2) nanoflakes as inherently electroactive labels for DNA hybridization detection. *Nanoscale*, 2014, *vol.* 6(20), 11971–11975.

[123] Mayorga–Martinez, C. C.; Ambrosi, A.; Eng, A. Y. S.; Sofer, Z.; Pumera, M. Metallic 1T–WS_2 for selective impedimetric vapor sensing. *Adv. Funct. Mater.*, 2015, *vol.* 25(35), 5611–5616.

[124] Liu, Y. M.; Zhou, M.; Liu, Y. Y.; Shi, G. F.; Zhang, J. J.; Cao, J. T.; Huang, K. J.; Chen, Y. H. Fabrication of electrochemiluminescence aptasensor based on in situ growth of gold nanoparticles on layered molybdenum disulfide for sensitive detection of platelet–derived growth factor–BB. *RSC Adv.*, 2014, *vol.* 4(44), 23888–22893.

[125] Shi, G. F.; Cao, J. T.; Zhang, J. J.; Liu, Y. M.; Chen, Y. H.; Ren, S. W. An electrochemiluminescence aptasensor based on flowerlike CdS–MoS_2 composites and DNAzyme for detection ofimmunoglobulin E. *Sens. Actuators, B*, 2015, *vol.* 220, 340–346.

[126] Wang, X.; Deng, W. P.; Shen, L.; Yan, M.; Ge, S. G.; Yu, J. H. A sensitive quenched electrochemiluminescent DNA sensor based on the catalytic activity of gold nanoparticle functionalized MoS_2. *New J. Chem.*, 2015, *vol.* 39(10), 8100–8107.
[127] Ou, X.; Tan, X. R.; Liu, X. F.; Chen, H. M.; Fan, Y.; Chen, S. H.; Wei, S. P. A cathodic luminol–based electrochemiluminescence biosensor for detecting cholesterol using 3D–MoS_2–PANI nanoflowers and Ag nanocubes for signal enhancement. *RSC Adv.*, 2015, *vol.* 5(81), 66409–66415.
[128] Ma, W. G.; Wang, L. N.; Zhang, N.; Han, D. X.; Dong, X. D.; Niu, L. Biomolecule–free, selective detection of o–diphenol and its derivatives with WS_2/TiO_2–based photoelectrochemical platform. *Anal. Chem.*, 2015, *vol.* 87(9), 4844–4850.
[129] Zang, Y.; Lei, J. P.; Hao, Q.; Ju, H. X. CdS/MoS_2 heterojunction–based photoelectrochemical DNA biosensor via enhanced chemiluminescence excitation. *Biosens. Bioelectron.*, 2016, *vol.* 77, 557–564.
[130] Li, H.; Yin, Z. Y.; He, Q. Y.; Li, H.; Huang, X.; Lu, G.; Fam, D. W. H.; Tok, A. L.; Y.; Zhang, Q.; Zhang, H. Fabrication of single– and multilayer MoS_2 film–based field–effect transistors for sensing NO at room temperature. *Small*, 2012, *vol.* 8(1), 63–67.
[131] He, Q. Y.; Zeng, Z. Y.; Yin, Z. Y.; Li, H.; Wu, S. X.; Huang, X.; Zhang, H. Fabrication of flexible MoS_2 thin–film transistor arrays for practical gas–sensing applications. *Small*, 2012, *vol.* 8(19), 2994–2999.
[132] Late, D. J.; Huang, Y. K.; Liu, B.; Acharya, J.; Shirodkar, S. N.; Luo, J. J.; Yan, A.; Charles, D.; Waghmare, U. V.; Dravid, V. P.; Rao, C. N. R. Sensing behavior of atomically thin–layered MoS_2 transistors. *ACS Nano*, 2013, *vol.* 7(6), 4879–4891.
[133] Wang, L.; Wang, Y.; Wong, J. I.; Palacios, T.; Kong, J.; Yang, H. Y. Functionalized MoS_2 nanosheet–based field–effect biosensor for label–free sensitive detection of cancer marker proteins in solution. *Small*, 2014, *vol.* 10(6), 1101–1105.
[134] Sarkar, D.; Liu, W.; Xie, X. J.; Anselmo, A. C.; Mitragotri, S.; Banerjee, K. MoS_2 field–effect transistor for next–generation label–free biosensors. *ACS Nano*, 2014, *vol.* 8(4), 3992–4003.
[135] Sarkar, D.; Xie, X. J.; Kang, J. H.; Zhang, H. J.; Liu, W.; Navarrete, J.; Moskovits, M.; Banerjee, K. Functionalization of transition metal dichalcogenides with metallic nanoparticles: implications for doping and gas–sensing. *Nano Lett.*, 2015, *vol.* 15(5), 2852–2862.

[136] Liu, B. L.; Chen, L.; Liu, G.; Abbas, A. N.; Fathi, M.; Zhou, C. W. High–performance chemical sensing using schottky–contacted chemical vapor deposition grown monolayer MoS_2 transistors. *ACS Nano*, 2014, *vol.* 8(5), 5304−5314.

[137] Huo, N. J.; Yang, S. X.; Wei, Z. M.; Li, S. S.; Xia, J. B.; Li, J. B. Photoresponsive and gas sensing field–effect transistors based on multilayer WS_2 nanoflakes. *Sci. Rep.*, 2014, *vol.* 4, 5209.

[138] Late, D. J.; Doneux, T.; Bougouma, M. Single–layer $MoSe_2$ based NH_3 gas sensor. *Appl. Phys. Lett.*, 2014, *vol.* 105(23), 233103.

[139] Xie, T.; Xie, G. Z.; Su, Y. J.; Du, H. F.; Ye, Z. B.; Jiang, Y. D. Ammonia gas sensors based on poly(3–hexylthiophene)–molybdenum disulfide film transistors. *Nanotechnology*, 2016, *vol.* 27(6), 065502.

[140] Friedman, A. L.; Perkins, F. K.; Cobas, E.; Jernigan, G. G.; Campbell, P. M.; Hanbicki, A. T.; Jonker, B. T. Chemical vapor sensing of two–dimensional MoS_2 field effect transistor devices. *Solid–State Electron.*, 2014, *vol.* 101(S1), 2−7.

[141] Lee, J.; Dak, P.; Lee, Y.; Park, Y.; Choi, W.; Alam, M. A.; Kim, S. Two–dimensional layered MoS_2 biosensors enable highly sensitive detection of biomolecules. *Sci. Rep.*, 2014, *vol.* 4, 7352.

[142] Kuru, C.; Choi, C.; Kargar, A.; Choi, D.; Kim, Y. J.; Liu, C. H.; Yavuz, S.; Jin, S. MoS_2 nanosheet–Pd nanoparticle composite for highly sensitive room temperature detection of hydrogen. *Adv. Sci.*, 2015, *vol.* 2(4), 1500004.

[143] Samnakay, R.; Jiang, C.; Rumyantsev, S. L.; Shur, M. S.; Balandin, A. A. Selective chemical vapor sensing with few–layer MoS_2 thin–film transistors: comparison with graphene devices. *Appl. Phys. Lett.*, 2015, *vol.* 106(2), 023115.

[144] Nam, H.; Oh, B. R.; Chen, P. Y.; Chen, M.; Wi, S.; Wan, W. J.; Kurabayashi, K.; Liang, X. G. Multiple MoS_2 transistors for sensing molecule interaction kinetics. *Sci. Rep.*, 2015, *vol.* 5, 10546.

[145] Nam, H.; Oh, B. R.; Chen, P. Y.; Yoon, J. S.; Wi, S.; Chen, M.; Kurabayashi, K.; Liang, X. G. Two different device physics principles for operating MoS_2 transistor biosensors with femtomolar–level detection limits. *Appl. Phys. Lett.*, 2015, *vol.* 107(1), 012105.

[146] Cho, B.; Yoon, J.; Lim, S. K.; Kim, A. R.; Kim, D. H.; Park, S. G.; Kwon, J. D.; Lee, Y. J.; Lee, K. H.; Lee, B. H.; Ko, H. C.; Hahm, M. G. Chemical sensing of 2D graphene/MoS_2 heterostructure device. *ACS Appl. Mater. Interfaces*, 2015, *vol.* 7, 16775−16780.

[147] Zhang, S. L.; Yue, H. Y.; Liang, X. S.; Yang, W. C. Liquid–phase co–exfoliated graphene/MoS_2 nanocomposite for methanol gas sensing. *J. Nanosci. Nanotechnol.*, 2015, *vol.* 15(10), 8004–8009.

[148] Lin, Y. F.; Xu, Y.; Lin, C. Y.; Suen, Y. W.; Yamamoto, M.; Nakaharai, S.; Ueno, K.; Tsukagoshi, K. Origin of noise in layered $MoTe_2$ transistors and its possible use for environmental sensors. *Adv. Mater.*, 2015, *vol.* 27(42), 6612–6619.

[149] Tong, Y.; Lin, Z. H.; Thong, J. T. L.; Chan, D. S. H.; Zhu, C. X. MoS_2 oxygen sensor with gate voltage stress induced performance enhancement. *Appl. Phys. Lett.*, 2015, *vol.* 107(12), 123105.

[150] Donarelli, M.; Prezioso, S.; Perrozzi, F.; Bisti, F.; Nardone, M.; Giancaterini, L.; Cantalini, C.; Ottaviano, L. Response to NO_2 and other gases of resistive chemically exfoliated MoS_2–based gas sensors. *Sens. Actuators, A*, 2015, *vol.* 207, 602–613.

[151] Jiang, S.; Cheng, R.; Ng, R.; Huang, Y.; Duan, X. F. Highly sensitive detection of mercury(II) ions with few–layer molybdenum disulfide. *Nano Res.*, 2015, *vol.* 8(1), 257–262.

[152] Lee, D. W.; Lee, J.; Sohn, I. Y.; Kim, B. Y.; Son, Y. M.; Bark, H.; Jung, J.; Choi, M.; Kim, T. H.; Lee, C.; Lee, N. E. Field–effect transistor with a chemically synthesized MoS_2 sensing channel for label–free and highly sensitive electrical detection of DNA hybridization. *Nano Res.*, 2015, *vol.* 8(7), 2340–2350.

[153] Pawbake, A. S.; Waykar, R. G.; Late, D. J.; Jadkar, S. R. Highly transparent wafer–scale synthesis of crystalline WS_2 nanoparticle thin film for photodetector and humidity–sensing applications. *ACS Appl. Mater. Interfaces*, 2016, *vol.* 8(5), 3359–3365.

[154] Mustafa, G.; Hussain, M.; Iqbal, N.; Dickert, F. L.; Lieberzeit, P. A. Quartz crystal microbalance sensor based on affinity interactions between organic thiols and molybdenum disulfide nanoparticles. *Sens. Actuators, B*, 2012, *vol.* 162(1), 63–67.

[155] Bajwa, S. Z.; Mustafa, G.; Samardzic, R.; Wangchareansak, T.; Lieberzeit, P. A. Nanostructured materials with biomimetic recognition abilities for chemical sensing. *Nanoscale Res. Lett.*, 2012, *vol.* 7(1), 328.

[156] Duerloo, K. A. N.; Ong, M. T.; Reed, E. J. Intrinsic piezoelectricity in two–dimensional materials. *J. Phys. Chem. Lett.*, 2012, *vol.* 3(19), 2871–2876.

[157] Zhu, H. Y.; Wang, Y.; Xiao, J.; Liu, M.; Xiong, S. M.; Wong, Z. J.; Ye, Z. L.; Ye, Y.; Yin, X. B.; Zhang, X. Observation of piezoelectricity in free–standing monolayer MoS_2. *Nat. Nanotechnol.*, 2015, *vol.* 10(2), 151−155.

[158] Park, W.; Yang, J. H.; Kang, C. G.; Lee, Y. G.; Hwang, H. J.; Cho, C.; Lim, S. K.; Kang, S. C.; Hong, W. K.; Lee, S. K.; Lee, S.; Lee, B. H. Characteristics of a pressure sensitive touch sensor using a piezoelectric PVDF–TrFE/MoS_2 stack. *Nanotechnology*, 2013, *vol.* 24(47), 475501.

[159] Qi, J. J.; Lan, Y. W.; Stieg, A. Z.; Chen, J. H.; Zhong, Y. L.; Li, L. J.; Chen, C. D.; Zhang, Y.; Wang, K. L. Piezoelectric effect in chemical vapour deposition–grown atomic–monolayer triangular molybdenum disulfide piezotronics. *Nat. Commun.*, 2015, *vol.* 6, 7430.

[160] Yu, S.; Eshun, K.; Zhu, H.; Li, Q. L. Novel two–dimensional mechano–electric generators and sensors based on transition metal dichalcogenides. *Sci. Rep.*, 2015, *vol.* 5, 12854.

Biographical Sketch

Yong Wang

Affiliation: College of Chemistry, Nanchang University, Nanchang, China

Education:

2008 – 2012	Ph. D. Degree, Analytical Chemistry, State Key Laboratory of Electroanalytical Chemistry, Changchun institute of Applied Chemistry, Chinese Academy of Science (CAS) Supervised by Prof. Xiurong Yang (Academician of CAS)
2005 – 2008	M.S. Degree, Analytical Chemistry, Department of Chemistry, Nanchang University Supervised by Prof. Yongnian Ni
2001 – 2005	B.S. Degree, Chemistry, Department of Chemistry, Nanchang University Supervised by Prof. Yongnian Ni

Business Address: 999 Xuefu Road, HongGuTang Newly-Developed Area, Nanchang, Jiangxi 330031, P. R. China

Research and Professional Experience: Synthesis, characterization and sensing applications of nanomaterials, Applied chemometrics

Publications Last 3 Years:

1. Yong Wang*, Jie Hu, Qianfen Zhuang, Yongnian Ni*, Enhancing sensitivity and selectivity in a label-free colorimetric sensor for detection of iron(II) ions with luminescent molybdenum disulfide nanosheet-based peroxidase mimetics, *Biosens. Bioelectron.*, 2016, 80, 111–117.
2. Yong Wang*, Jie Hu, Qianfen Zhuang, Yongnian Ni*, Label-free fluorescence sensing of lead(II) ions and sulfide ions based on luminescent molybdenum disulfide nanosheets, *ACS Sustainable Chem. Eng.*, 2016, 4(5), 2535–2541.
3. Jie Hu, Qianfen Zhuang, Yong Wang*, Yongnian Ni*, Label-free fluorescent catalytic biosensor for highly sensitive and selective detection of the ferrous ion in water samples using a layered molybdenum disulfide nanozyme coupled with an advanced chemometric model, *Analyst.*, 2016, 141(5), 1822–1829.
4. Qianfen Zhuang, Yong Wang*, Yongnian Ni*, Solid–phase synthesis of graphene quantum dots from the food additive citric acid under microwave irradiation and their use in live–cell imaging, *Luminescence*, 2016, 31(3), 746–753.
5. Yong Wang*, Qianfen Zhuang, Yongnian Ni*, Facile microwave–assisted solid–phase synthesis of highly fluorescent nitrogen–sulfur–codoped carbon quantum dots for cellular imaging applications, *Chem. Eur. J.*, 2015, 21(37), 13004–13011.
6. Yong Wang, Yao Gu, Yongnian Ni*, Serge Kokot, Inhibition effect of graphene oxide on the catalytic activity of acetylcholinesterase enzyme, *Luminescence*, 2015, 30(7), 940–946.
7. Yong Wang, Qianfen Zhuang, Yongnian Ni*, Fabrication of riboflavin electrochemical sensor based on homoadenine single-stranded DNA/molybdenum disulfide–graphene nanocomposite modified gold electrode, *J. Electroanal. Chem.*, 2015, 736, 47–54.
8. Yong Wang, Yongnian Ni*, Molybdenum disulfide quantum dots as a photoluminescence sensing platform for 2,4,6–trinitrophenol detection, *Anal. Chem.*, 2014, 86(15), 7463–7470.
9. Yong Wang, Minghua Mei, Yongnian Ni*, Serge Kokot, Combined NIR/MIR analysis: A novel method for the classification of complex substances such as *Illicium verum Hook. F.* and its adulterants, *Spectrochim. Acta, Part A*, 2014, 130, 539–545.
10. Yong Wang, Ruirui Zhu, Yongnian Ni*, Serge Kokot, Competitive interactions of anti–carcinogens with serum albumin: A spectroscopic study

of bendamustine and dexamethasone with the aid of chemometrics, *Spectrochim. Acta, Part A*, 2014, 123, 241–248.

11. Yong Wang, Yongnian Ni*, Combination of UV–Vis spectroscopy and chemometrics to understand protein–nanomaterial conjugate: A case study on human serum albumin and gold nanoparticles, *Talanta*, 2014, 119, 320–330.
12 Yong Wang, Yongnian Ni*, New insight into protein–nanomaterial interactions with UV–visible spectroscopy and chemometrics: human serum albumin and silver nanoparticles, *Analyst*, 2014, 139(2), 416–424.
13. Yong Wang, Baohui Li, Yongnian Ni*, Serge Kokot, Multi–wavelength high–performance liquid chromatography: An improved method for analysis of complex substances such as Radix Paeoniae herbs, *Chemom. Intell. Lab. Syst.*, 2014, 130, 159–165.
14. Yong Wang, Baohui Li, Yongnian Ni*, Serge Kokot, A high–performance liquid chromatography method for quality control of complex substances: analysis of marker alkaloids and flavonoids of the herb Herba leonuri (Yimucao), *Anal. Methods.*, 2013, 5(21), 6051–6057.
15. Xinrong Guo, Yong Wang, Fangying Wu, Yongnian Ni*, Serge Kokot, Preparation of protonated, two-dimensional graphitic carbon nitride nanosheets by exfoliation, and their application as a fluorescent probe for trace analysis of copper(II), *Microchim. Acta*, 2016, 183(2), 773–780.
16. Xinrong Guo, Yong Wang, Fangying Wu, Yongnian Ni*, Serge Kokot, The use of tungsten disulfide dots as highly selective, fluorescent probes for analysis of nitrofurazone, *Talanta*, 2015, 144, 1036–1043.
17. Zhuo Li, Yong Wang, Yongnian Ni*, Serge Kokot, Fluorescence analysis of 6–mercaptopurine with the use of a nano–composite consisting of BSA–capped Au nano–clusters and core–shell Fe_3O_4–SiO_2 nanoparticles, *Biosens. Bioelectron.*, 2015, 70, 246–253.
18. Zhuo Li, Yong Wang, Yongnian Ni*, Serge Kokot, A sensor based on blue luminescent graphene quantum dots for analysis of a common explosive substance and an industrial intermediate, 2,4,6–trinitrophenol, *Spectrochim. Acta, Part A*, 2015, 137, 1213–1221.
19. Zhuo Li, Yong Wang, Yongnian Ni*, Serge Kokot, A colorimetric method of analysis for trace amounts of hydrogen peroxide with the use of the nano–properties of molybdenum disulfide, *Analyst*, 2015, 140(4), 1119–1126.
20 Zhuo Li, Yong Wang, Yongnian Ni*, Serge Kokot, A rapid and label–free dual detection of Hg (II) and cysteine with the use of fluorescence switching of graphene quantum dots, *Sens. Actuators. B*, 2015, 207, 490–497.

21. Ruiling Sun, Yong Wang, Yongnian Ni*, Serge Kokot, Graphene quantum dots and the resonance light scattering technique for trace analysis of phenol in different water samples, *Talanta*, 2014, 125, 341–346.
22. Qiulan Zhang, Yong Wang, Yongnian Ni*, Serge Kokot, Analysis of complex molecular systems: the impact of multivariate analysis for resolving the interactions of small molecules with biopolymers – a review, *Anal. Lett.*, 2014, 47(7), 1089–1106.
23. Zhuo Li, Yong Wang, Yongnian Ni*, Serge Kokot, Unmodified silver nanoparticles for rapid analysis of the organophosphorus pesticide, dipterex, often found in different waters, *Sens. Actuators. B*, 2014, 193, 205–211.
24. Ruiling Sun, Yong Wang, Yongnian Ni*, Serge Kokot, Simultaneous kinetic spectrometric determination of three flavonoid antioxidants in fruit with the aid of chemometrics, *Spectrochim. Acta, Part A*, 2014, 122, 529–535.
25. Ruiling Sun, Yong Wang, Yongnian Ni*, Serge Kokot, Spectrophotometric analysis of phenols, which involves a hemin–graphene hybrid nanoparticles with peroxidase–like activity, *J. Hazard. Mater.*, 2014, 266, 60–67.

In: Nanosheets and Nanospheres
Editor: Deborah Wright
ISBN: 978-1-53610-203-1

Chapter 3

Fabrication and Function of "Polymer Nanosphere Multilayered Organization"

Manami Iizuka and Atsuhiro Fujimori *

Graduate School of Science and Engineering,
Saitama University, Sakura-ku, Saitama, Japan

Abstract

This chapter introduces on a new functionality exhibited by "polymer nanosphere multilayered organization," a new type of molecular organization, and the relationship between their structure and function. The polymer nanosphere multilayered organization is a fine structural material formed by the accumulation of single-particle layers of a hydrophobic polymer at the air/water interface; these single-particle layers have uniform height along the *c*-axis. We constructed a multiparticle layered organization of aromatic polyamides with rigid main chains and flexible side chains by the Langmuir–Blodgett (LB) technique, which resulted in a highly regular arrangement along the *c*-axis. The particle arrangement was estimated by performing out-of-plane X-ray diffraction (XRD) analysis and atomic force microscopic (AFM) observation. The results suggest that a double-particle layered structure (Y-type) is formed by the LB technique,

* Corresponding author: Tel. and Fax: +81-48-858-3503; E-mail address: fujimori@fms.saitama-u.ac.jp (A. Fujimori).

forming amphiphilic particles at the air/water interface. Copolymers with highly hydrophobic carbazole contents and both hydrogenated and fluorinated side-chains also formed a single-particle layer at the air/water interface and exhibited multiparticle layers by a LB technique. Therefore, it is possible to control the formation of single- and double-particle layered structure using these techniques. Further, it was found that multiparticle layered organization of polymer nanospheres and polymer nanosheets could be formed simultaneously with the same component material. By employing the "alternate compression–relaxation method," high-density, low defect particle layers are formed with a clear increase in their crystallite sizes. In the case of a ternary comb copolymer containing a carbazole ring, one particle is formed by the assembly of approximately 60 units of collapsed monolayer-like double layers. This structure is stabilized by the formation of side-chain crystals in the interlayer, with oriented π–π stacking of carbazole rings, resulting in enhanced fluorescence emission intensity.

1. Fabrication and Structure of "Polymer Nanosphere Multilayered Organization"

1.1. Introduction

Materials with a layered organization such as graphite, aluminosilicate, biological membranes, cell membranes, Langmuir–Blodgett (LB) films [1–5], β-sheets in proteins [6–8], and lamella structures [9–11] of crystalline polymers are regarded as key materials in the organic/inorganic and biochemistry fields. Exfoliation [12–14], intercalation [15–17], deintercalation [18–20], electron transfer between layers [21–23], and so on are representative functions of this type of layered organization, which has thus attracted considerable interest in fundamental science.

By collapsing a monolayer consisting of organic polymer nanoparticles on the surface of water, it is possible to produce new molecular assemblies [24]. Figure 1(a) shows single-particle and multiparticle layers of hydrophobic polymers on a water surface fabricated by the LB technique. These nanoparticles form a layered organization by the bottom-up method. These multilayered polymer nanospheres are similar to colloidal crystals [25]. Colloidal particles are small objects with at least one characteristic dimension in the range of 1 nm to 1 μm. They have long been used as major components of various industrial products such as foods, inks, paints, toners, coatings, papers, cosmetics, photographic films, and rheological fluids. Further, colloidal crystals, which comprise three-dimensional periodic arrays of submicron particles, have attracted considerable attention because of their novel photonic applications in the form of photonic crystals. They have also been extensively studied in the context of materials science, chemistry, biology, condensed matter physics, applied optics, and fluid dynamics. Generally, it is well known that photonic crystals are nanostructures that exhibit periodic changes of refractive index. Figure 1(b) shows that the construction of a multiparticle layered organization such as that found in photonic crystals is possible by using the LB method. The LB method can alternately transfer particles with different refractive indices, similar to the formation of hetero film. However, colloidal crystals are generally composed of microsized particles and the particles are frequently expensive metals or inorganic materials.

Polymer nanospheres are a viable alternative to colloidal crystals. Figure 2 shows the formation, structure, and function of multilayered polymer nanospheres of aromatic polyamides. The atomic force microscopic (AFM) image shows the formation of a single-particle layer of an aromatic polyamide with a pentyl side chain ($PABA_5$) [26, 27] [Figure 2(a)]. The out-of-plane XRD profile of the LB-fabricated multiparticle layers of $PABA_5$ indicates higher-order reflection of the (00*l*) plane [Figure 2(b)]. By applying of three-dimensional paracrystal analysis [28], a *g* value, which is a measure of crystal distortion, of 0.52 was calculated from the result of out-of-plane XRD of this LB multiparticle layers [Figure 2(c)]. The interference colors resulting from the difference in layer thickness of the step-wise multiparticle layers of $PABA_5$ are easily visible [Figure 2(d)]. In addition, it is expected that fabrication of inexpensive, nanometer-size, soft material multiparticle layers by hybridization of particles with different refractive indices. This layered organization may show wave number selectivity at several tens of angstroms, such as in the region of vacuum ultraviolet or far-ultraviolet light [Figure 2(e)]. Further, surface properties of polymers are important for understanding adhesion, friction, and wear-resistance characteristics [29–32]. For example, wettability of LB films is sensitive to the flatness or roughness of these surfaces. It is expected application of multiparticle layers for various devices.

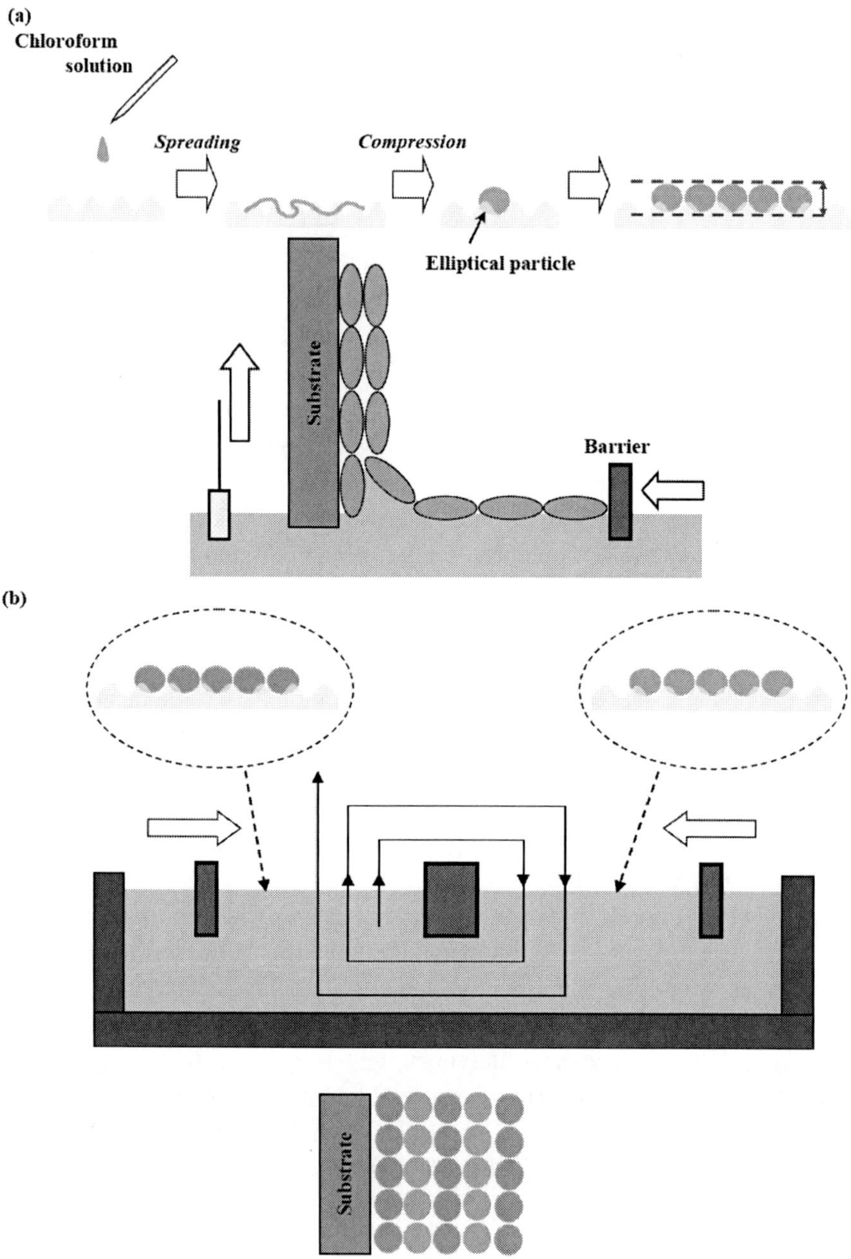

Figure 1. Formation mechanism of (a) single-particle layer on the water surface of hydrophobic polymers and (b) multiparticle layers by the LB technique.

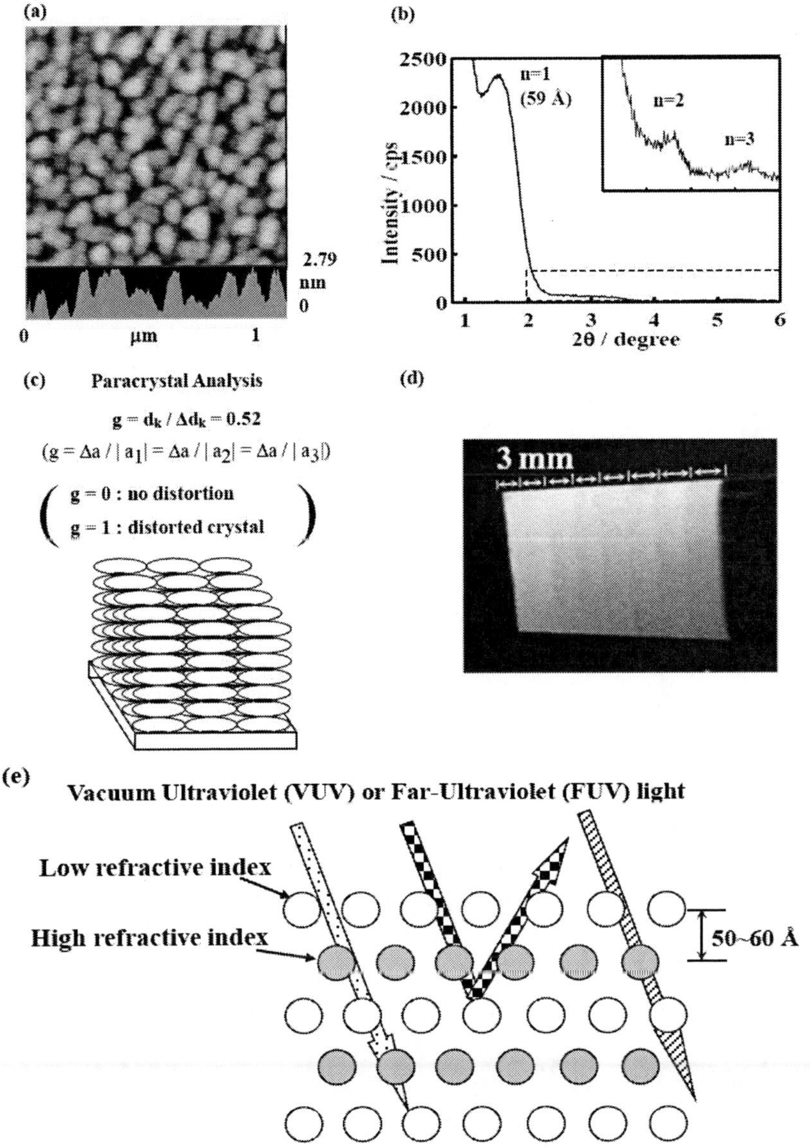

Figure 2. Formation, structure, and function of polymer nanosphere multilayered organization of aromatic polyamides: (a) AFM image of single-particle layer of poly-(*N*-alkylated benzamides) ($PABA_5$); (b) Out-of-plane XRD profile of LB multiparticle layers of $PABA_5$; (c) Paracrystal analysis of the LB multiparticle layers of $PABA_5$ and their illustration; (d) Appearance of structural color of step-wise multiparticle layers of $PABA_5$; and (e) Schematic illustration of wave number selectivity of multiparticle layered organization of polymer nanosphere at several tens of angstrom.

In a previous study, aromatic polyamides with a rigid main chain and flexible side chains of various lengths [poly-(*N*-alkylated benzamide), *abbrev.* PABA$_n$ *n*: number of carbon side chains] were synthesized by direct condensation polymerization [33, 34]. In the case of using of method of monolayer on the water surface of these polyamides, the surface morphology of PABA$_n$ systematically varied depending on the side-chain length [35–39]. In the first section of this chapter, the formation mechanism and functionality of the multiparticle layered organization of aromatic polyamides and other hydrophobic polymers were investigated by surface pressure–area (π–A) isotherms and AFM observation. These hydrophobic polymers correspond to ternary comb copolymers containing hydrophobic *N*-vinyl carbazole (NVCz) units. Previously, we constructed novel polymer nanosheets with a highly ordered layer structure with amorphous side chains and π–π stacking of NVCz in the form of ternary comb copolymers [40]. Therefore, the formation behavior of particles of these type ternary comb copolymers with highly hydrophobic NVCz content was also used. In addition, organo-modified zirconia and clay as particles of different refractive indices were used. In this section, multiparticle layered organization consisting of such particles was prepared and that's structure was investigated.

1.2. Materials

The PABA$_n$ compounds used in this study were synthesized by the direct condensation polymerization of *N*-alkylated aminobenzoic acid from methyl to heptadecyl substituents (n = 1–17) [26]. The average molecular weights, M_w, of these aromatic polyamides were 20,000–90,000 by gel permeation chromatography (GPC, JASCO-860-CO) measurement. The ternary comb copolymers used in this study were obtained by the copolymerization of NVCz with octadecyl acrylate (OA) and 2-(perfluorodecyl)ethyl acrylate ($FF_{10}EA$) at various monomer ratios (Figure 3). Copolymerization was carried out in an acetone solution at 50°C for 48 h using azobisisobutyronitrile (AIBN) as an initiator. These monomers and the initiator were purchased from Tokyo Kasei Co. Ltd. and Daikin Fine Chemicals Co. Ltd., respectively, and were used without further purification. The precipitated polymers were washed with acetone until they were free of monomers. The ternary comb copolymer compositions were determined by ^{1}H NMR (EX270 NMR, Nihon Densi Co. Ltd.) spectroscopy. The tacticity of the fluorinated homopolymer obtained by ^{1}H NMR analysis according to the reference was found to be almost syndiotactic (diad: 58%). In this study, we synthesized ternary copolymers with molecular weights of approximately 4.93×10^3 (molecular weight distribution, $M_w/M_n \approx 1.17$) estimated on the basis of GPC measurements.

According to the theory of the *Q–e* scheme proposed by Alfrey and Price [41], these ternary comb polymers form alternating copolymers of NVCz and long-chain acrylates. In this case, the *e* values of NVCz, OA, and $FF_{10}EA$ are −1.40, +1.12, and +0.66 respectively. Thus, it appears that the NVCz:long-chain acrylate copolymers form almost ideal alternating copolymers, especially for the 2:1:1 monomer ratio.

For the organo-modification of the inorganic particles analyzed in this study, a dispersed mixed solution was prepared by mixing a ZrO_2 aqueous solution and a methanol solution that contained long-chain carboxylic acids (e.g., stearic acid). Toluene was then poured into the dispersed solution with stirring [42], which caused the organo-modified ZrO_2 to migrate from the dispersion phase to the toluene phase. Next, the water, methanol, and any unreacted agent were removed from the reaction system. As a result, hydroxyl groups on the particle surface were exchanged with long alkyl carboxylates. (*These procedures are shown as Figure 12 in detail later).

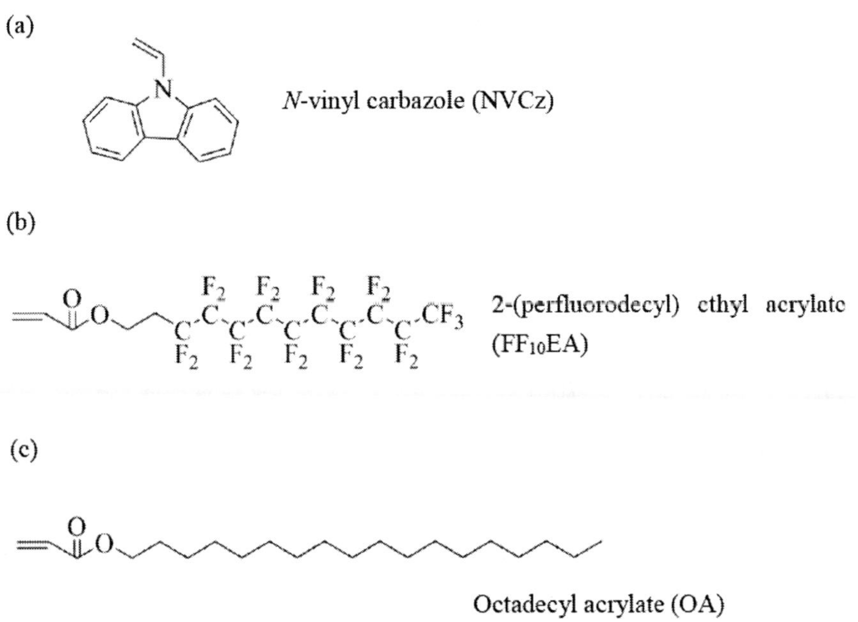

Figure 3. Chemical structure of (a) NVCz, (b) $FF_{10}EA$, and (c) OA.

Natural Na^+-montmorillonite with a cation-exchange capacity of 108.6 meq/100 g was supplied by Kunimine Co. Ltd. The organophilic clay was prepared using cation exchange reactions of natural clay (aqueous dispersion) with 0.1% aqueous solution of dimethyl dioctadecyl (DMDO) ammonium. Organo-clays with low surface coverage (commercial grade) [43] were formed in clay suspensions at 50°C by the addition of quaternary ammonium cations with stirring, whereas those with high surface coverage were formed at the oil/water interface (chloroform solutions of quaternary ammonium chlorides/clay suspensions) [44, 45]. (*These procedures are illustrated as Figure 14 in detail later. The high-surface-coverage modification method is similar to that of interfacial polycondensation. A schematic illustration of the cation-exchange reaction and the structure of the resulting organo-modified clay are also shown on lower side of Figure 14 in detail later.)

1.3. Procedures

1.3.1. Formation of Polymer Monolayers and Single-Particle Layers on Water Surface

The π–A isotherms were measured by a USI-3-22 film balance (USI Co. Ltd.) at 15°C. Monolayers and single-particle layers of aromatic polyamides prepared from chloroform solutions (approximately 10^{-4} M) were formed on distilled water (resistivity approximately 18 MΩ·cm). After the evaporation of the chloroform for 5 min, the π–A isotherms were recorded at a compression speed of 9.6 $cm^2 \cdot min^{-1}$. These aromatic polyamides formed highly condensed films. These films were transferred onto solid substrates at 15°C and an appropriate surface pressure (25 mNm^{-1}) to obtain alternating Y-type films by the LB method.

The monolayers and single-particle layers of the ternary comb copolymers with NVCz were spread from the chloroform or chloroform/trifluoroacetic acid [90/10 (v/v)] mixed solutions (approximately 10^{-4} M) onto the distilled water (resistivity approximately 18 MΩ·cm). After the evaporation of the solvent, the π–A isotherms were recorded at a compression speed of 9.6 $cm^2 \cdot min^{-1}$. These monolayers were transferred onto glass (XRD sample) and mica (AFM) substrates by the LB method to obtain alternating Y-type films at 15°C and an appropriate surface pressure (5 or 30 mNm^{-1}).

Monolayers of organo-modified ZrO_2 were formed by spreading a chloroform solution (approximately 10^{-4} M) onto distilled water (resistivity approximately 18 MΩ·cm). After the evaporation of the chloroform for 5 min, the π–A isotherms were recorded at a compression speed of 9.6 $cm^2 \cdot min^{-1}$. The air/water interface was maintained at a constant temperature of 15°C by circulating thermostatic water around the trough. These monolayers were transferred onto glass (XRD sample) and mica (AFM) substrates by the LB method.

1.3.2. Molecular Arrangement in the Films

The large spacing between the layer structures of the films transferred onto the glass substrates were measured using an out-of-plane X-ray diffractometer (Rigaku, Rint-Ultima III, Cu Kα radiation, 40 kV, 40 mA) equipped with a graphite monochromator. The in-plane spacing of the two-dimensional lattice of the films was determined using an X-ray diffractometer with different geometrical arrangements [46, 47] (Bruker AXS, MXP-BX, Cu Kα radiation, 40 kV, 40 mA, an instrument specially made to order) and equipped with a parabolic graded multilayer mirror. The X-rays were incident at an angle of 0.2° and the films were slow scanned at a speed of 0.05°/80 s. Thus, in-plane XRD measurements were carried out at a monomolecular resolution.

1.3.3. Observation of Surface Morphology in Organized Molecular Films

The surface morphologies of the transferred films were observed using a scanning probe microscope (SPA300 with SPI-3800 probe station, Seiko Instrument) and microfabricated rectangular Si cantilevers with integrated pyramidal tips by applying a constant force of 1.4 Nm^{-1}.

1.4. Formation of Single Particle Layer of Polymers at Air/Water Interface

Figure 4 shows the π–A isotherms and AFM images of monolayer and single-particle layer aromatic polyamides of $PABA_1$, $PABA_3$, $PABA_5$, $PABA_7$, and $PABA_{17}$. In an aromatic polyamide, the surface morphology of the monolayer depends on the length of the side chain. Carbon numbers of 1–5, 7, and 17 in the side chain form a single particle layer, a shapeless domain, and a giant domain, respectively [26]. Influence of side-chain length of the polyamide on particle formation is mentioned by reference number 26. In a previous study, it was confirmed the inside of polyamide particle is amorphous [26]. In addition, aromatic polyamide is well known as a heat-resistant material. It is expected these particle film has both durability and thermal stability.

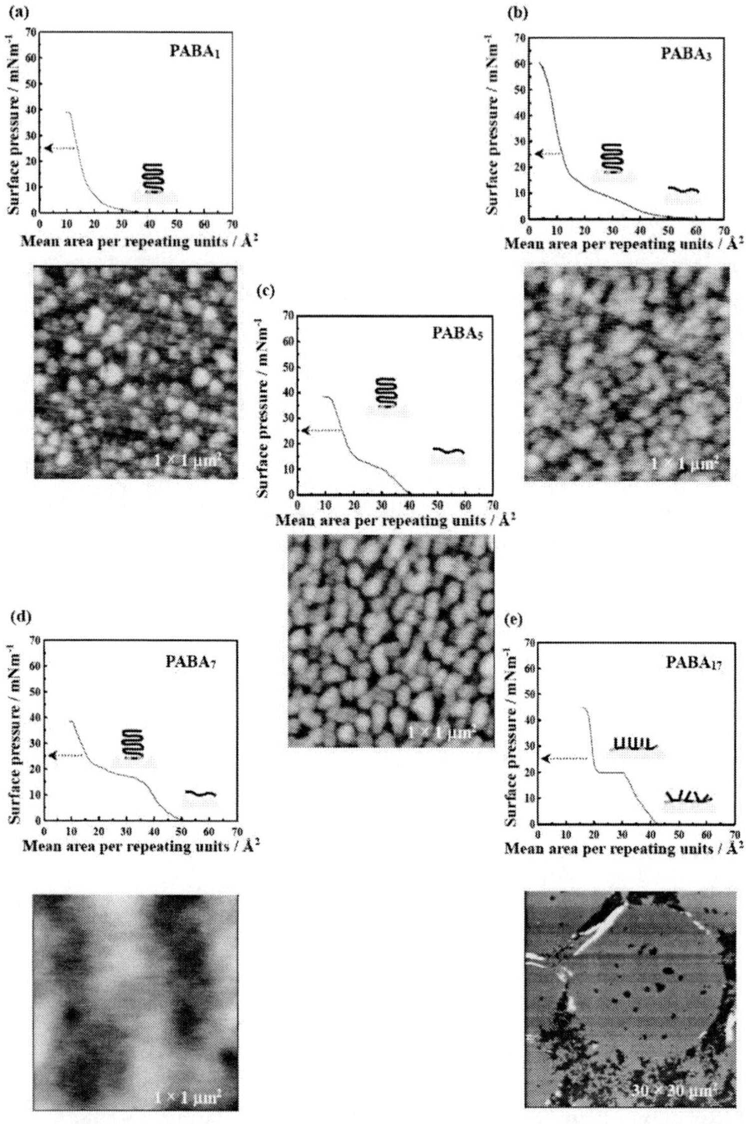

Figure 4. Surface pressure–area isotherms and AFM images of monolayers or single-particle layers of aromatic polyamides: (a) $PABA_1$, (b) $PABA_3$, (c) $PABA_5$, (d) $PABA_7$, and (e) $PABA_{17}$.

Figure 5 shows π–A isotherms and corresponding AFM images of $PABA_5$, along with schematic illustration of the formation mechanism of a single-particle layer. In this case, a monolayer film is formed in a low-surface-pressure region, whereas the main chain is folded due to compression to form a single-particle layer film in a high-surface-pressure region. In other words, the structural formation of the single-particle layer is induced by the collapse of the monolayer. In the case of particle of polyamide system, a particle is composed by about one hundred number of polymer chain.

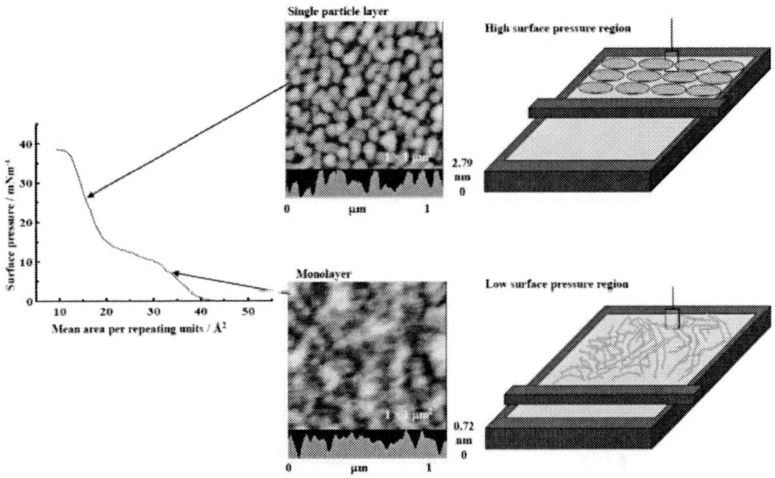

Figure 5. Surface pressure–area isotherms and corresponding AFM images of $PABA_5$ with schematic illustration of formation mechanism of single-particle layer.

1.5. Fabrication of "Polymer Nanosphere Multilayred Organization" as Multi-particle layer by Langmuir-Blodgett Technique

Figure 6 shows the dependence of the morphology of the $PABA_1$ and $PABA_7$ particle layers on the transferred subphase temperature. Figure 6(a) shows AFM images of the single-particle layers of $PABA_1$ at transferred subphase temperatures of 15°C and 35°C. It was found that particle formation was promoted with increasing subphase temperature. Figure 5(b) shows AFM images of the monolayer and single-particle layer of $PABA_7$ at transferred subphase temperatures of 15°C and 25°C. In this polymer, in addition to the formation of a shapeless domain at 15°C, a single-particle layer can also be formed by an increase in subphase temperature until reaching 25°C. In the case of other polyamide, dependence of surface morphology of monolayer on subphase temperature such as $PABA_1$ and $PABA_7$ was not observed. Figure 6(c) shows the out-of-plane XRD profiles of the LB multilayers and multiparticle layers of $PABA_7$ at transferred subphase temperatures of 15°C and 25°C. The layer period from the multilayers transferred at 15°C could not be confirmed, but it can be clearly seen in the multiparticle layers transferred at 25°C. At 25°C, a single-particle layer with a uniform height is formed. Particle formation at the air/water interface is accelerated by the increase in the mobility of the molecular chain with the subphase temperature rise.

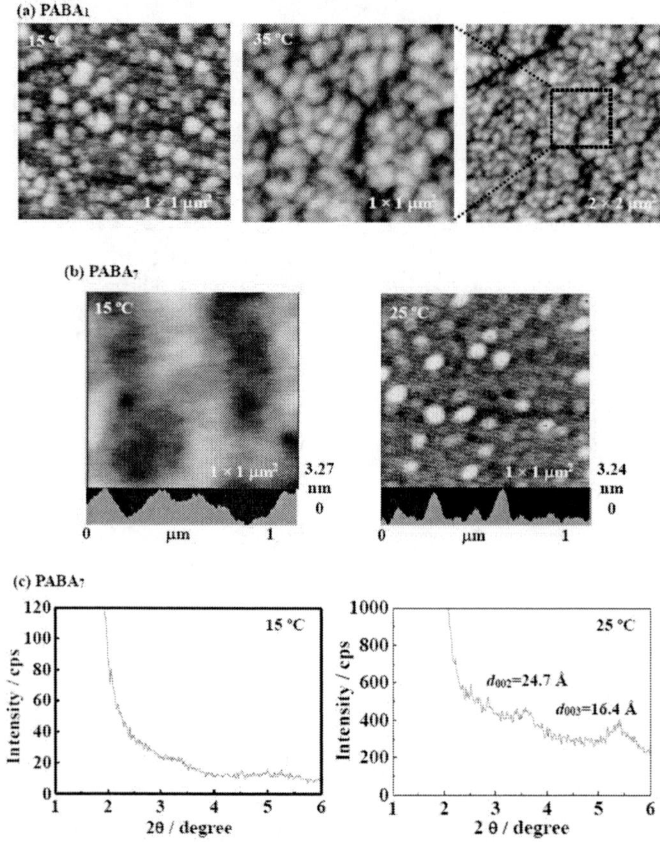

Figure 6. Dependence of morphology of $PABA_1$ and $PABA_7$ particle layers on transferred subphase temperature. (a) AFM images of single-particle layers of $PABA_1$ at 15 and 35°C; (b) AFM images of monolayer and single-particle layers of $PABA_7$ at 15 and 25°C; and (c) Out-of-plane XRD profiles of LB multilayers and multiparticle layers of $PABA_7$ at 15 and 25°C.

Figure 7 shows the formation of single-particle layers of other polymers. Figure 7(a) shows the surface pressure–area isotherm of a single-particle layer of *N*-vinylcarbazole on the water surface. The particle formation behavior at the air/water interface is confirmed in this carbazole system also. The poly-(*N*-vinylcarbazole) formed a condensed monolayer with an extremely limited area. Figure 7(b) shows an AFM image of the single-particle layer of *N*-vinylcarbazole. In this compound, the formation of a single-particle layer instead of a monolayer film can be seen, and the height of the particles is approximately 5 nm. Figure 7(c) shows the TEM image of a single-particle layer of polystyrene taken from Kumaki [48]. Hence, the experimental results of this system are close to those previously reported for a polystyrene system at the air/water interface. Figure 7(d) shows the out-of-plane XRD profile of the LB multiparticle layers of *N*-vinylcarbazole. In this case, a highly ordered particle-layer structure is constructed.

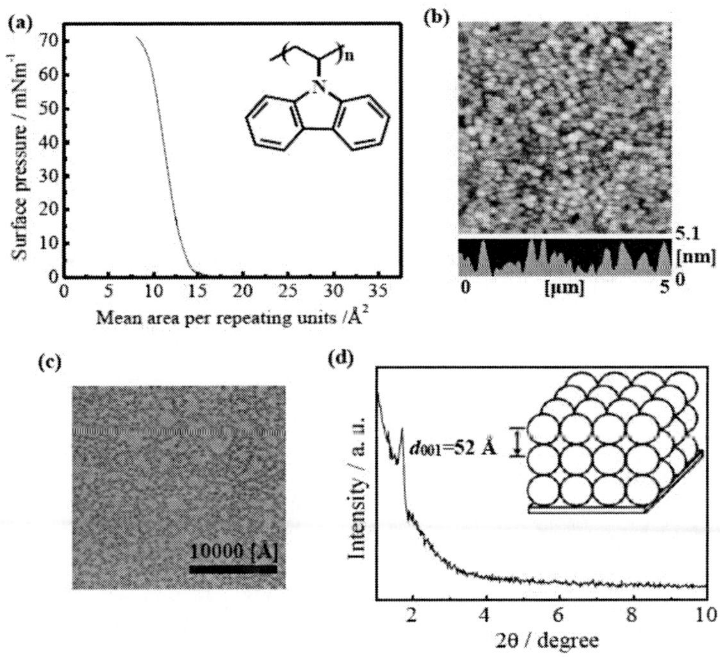

Figure 7. Formation of single-particle layers of other polymers. (a) Surface pressure–area isotherm of single-particle layer on the water surface of *N*-vinylcarbazole; (b) AFM image of single-particle layer of *N*-vinylcarbazole; (c) TEM image of single-particle layer of polystyrene taken from Kumaki [28]. (d) Out-of-plane XRD profiles of LB multiparticle layers of *N*-vinylcarbazole.

We previously reported on organized molecular films of ternary comb copolymers having carbazole groups, hydrocarbon chains, and fluorocarbon chains [40, 49, 50]. Figure 8 shows the formation and structure of a ternary comb copolymer nanosheet for NVCz:OA:FF_{10}EA = 2:1:1. Figure 8(a) shows the surface pressure–area isotherm of a monolayer on the water surface. This copolymer forms an extremely stable condensed film. Figure 8(b) shows the AFM images of the Z-type monolayer. Over the mesoscopic region, this monolayer shows a very smooth surface morphology. Figure 8(c) shows the in-plane XRD profile of the LB multilayers. Because an amorphous structure is formed in the film plane, diffraction does not occur. In other words, the amorphous phase is supported by the layer [51]. The layered structure is stable over a long period of time because the amorphous region is semi-permanently retained [52–54]. In addition, the main chain is firmly tied by π–π interactions between the carbazole rings. Figure 8(d) shows the out-of-plane XRD profile of LB multilayers of the NVCz:OA:FF_{10}EA = 2:1:1 copolymer. This multilayer film with a layer period of 48 Å is highly ordered.

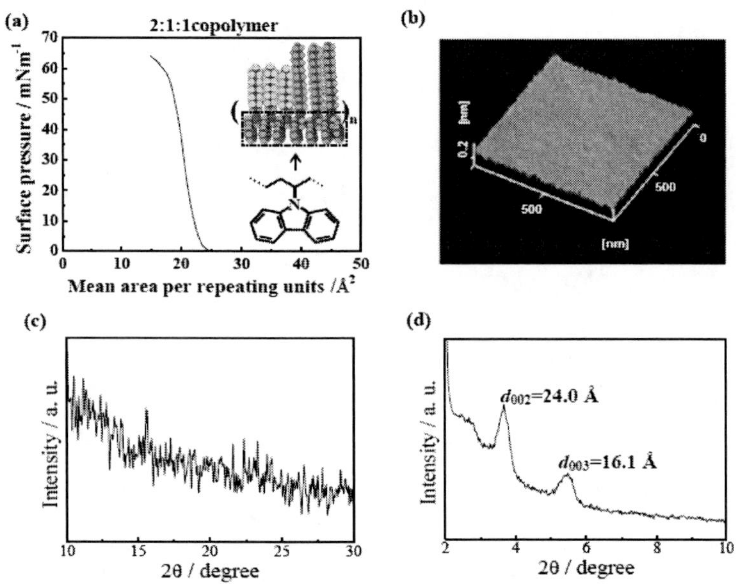

Figure 8. Formation and structure of polymer nanosheet of ternary comb copolymer for NVCz:OA:FF_{10}EA = 2:1:1. (a) Surface pressure–area isotherm of monolayer on the water surface at 15°C; (b) AFM images of Z-type monolayer; (c) In-plane XRD profile of LB multilayers; and (d) Out-of-plane XRD profile of LB multilayers of NVCz:OA:FF_{10}EA = 2:1:1 copolymer.

1.6. Elucidation of Formation Condition of Interfacial Single Particle Layer of Hydrophobic Polymers on the Water Surface

Carbazole rings are highly hydrophobic functional groups. In this ternary copolymer, a single-particle layer is expected to form with the increase in the ratio of carbazole rings. In order to confirm the dependence of the transition between monolayer and single-particle layer on the copolymerization ratio, we formed an interfacial molecular film on a ternary comb copolymer with a high carbazole ratio. Figures 9(a) and (b) shows π–A isotherms and AFM images, respectively, of Z-type monolayers and single-particle layers of NVCz:OA:$FF_{10}EA$ = 3:1:1, 4:1:1, and 6:1:1 copolymers. With increasing content of carbazole, a plateau based on a two-dimensional transition occurs. The morphology of the monolayer surface in the low-surface-pressure region and of the single-particle layer in the high-surface-pressure region was confirmed by AFM observation of the films transferred on a solid before and afterthe plateau. In the case of particle of carbazole system, a particle is composed by about seventy number of polymer chain.

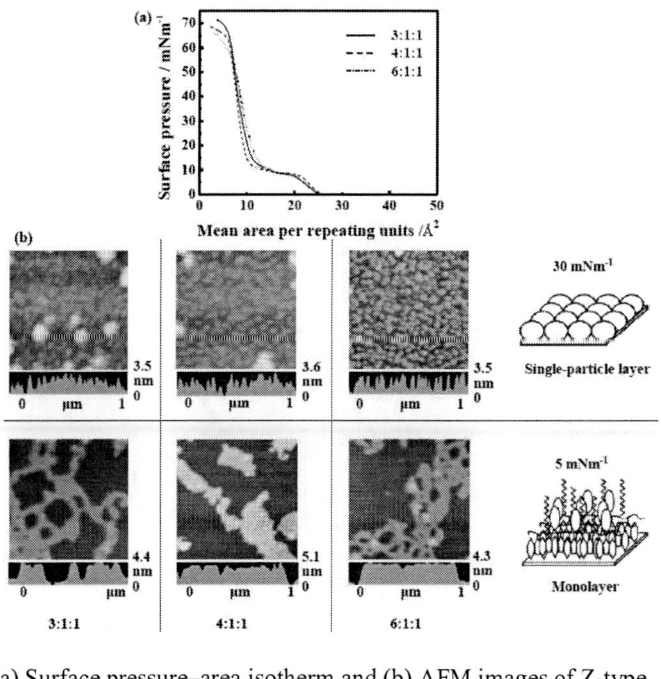

Figure 9. (a) Surface pressure–area isotherm and (b) AFM images of Z-type monolayers and single-particle layers of NVCz:OA:$FF_{10}EA$ = 3:1:1, 4:1:1, and 6:1:1 copolymers at 15°C.

Figure 10 shows the dependence of the morphology of a ternary comb copolymer with NVCz on subphase temperature. Figures 10(a), (b), and (c) show π–A isotherms on the water surface of monolayers and single-particle layers of NVCz:OA:$FF_{10}EA$ = 3:1:1, 4:1:1, and 6:1:1 at 10, 15, and 20°C, respectively. At 10°C, the value of limiting area of monolayer for these copolymers is reasonable as compared with their molecular area. However, the limiting area of these copolymers is too narrow at 20°C. That is to say, it is suggested that these copolymers are transformed to the monolayer and single-particle layer at the subphase temperature 10 and 20°C, respectively. At a subphase temperature of 15°C, the transition from the monolayer to the two-dimensional single-particle layer was confirmed. Figure 10(d) shows an AFM image of Z-type single-particle layers and monolayers of NVCz:OA:$FF_{10}EA$ = 4:1:1 copolymer transferred at subphase temperatures of 10°C and 20°C. As with the polyamide-based nanosphere system, particle formation is dependent on subphase temperature, and it was found that particle formation is impossible at 10°C, whereas particles formed immediately at 20°C. An acceleration of particle formation is thus due to the increased mobility of the molecular chain at high subphase temperatures.

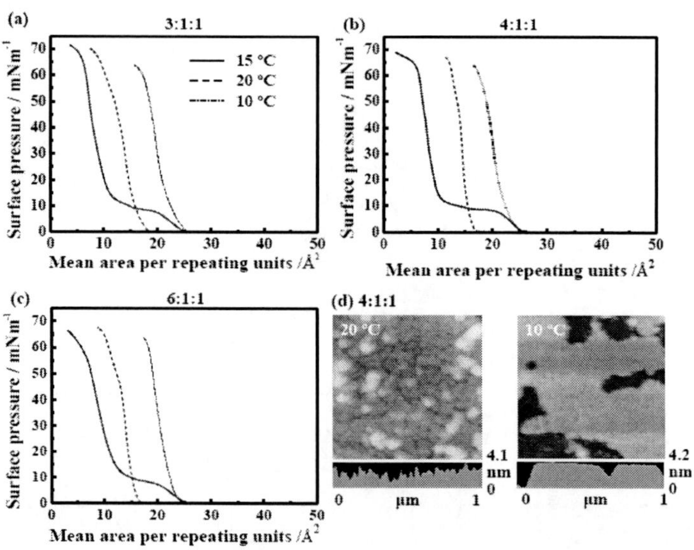

Figure 10. Dependence of morphology of ternary comb copolymer with NVCz on the subphase temperature. Surface pressure–area isotherms on the water surface of monolayers and single-particle layers of NVCz:OA:$FF_{10}EA$ = 3:1:1 (a), 4:1:1 (b), and 6:1:1 (c). (d) AFM images of Z-type single-particle layers and monolayers of NVCz:OA:$FF_{10}EA$ = 4:1:1 copolymer transferred at 20 and 10°C.

Figure 11 shows out-of-plane XRD profiles of LB multiparticle layers and multilayers of the NVCz:OA:FF_{10}EA = 4:1:1 copolymer, a schematic illustration of each layered structure, and the formation mechanism of a single-particle layer of NVCz:OA:FF_{10}EA copolymers on the water surface. With the compression of the monolayer at the air/water interface, the polymer chains folded. They are then stabilized by the outward orientation of the highly hydrophobic carbazole units in the high-surface-pressure region. As the result, a single-particle layer of uniform height is formed because all the polymer chains form a similar structure. In addition, a double-layered structure is formed in the case shown in Figure 11(a), whereas the single-layered structure is formed in the case shown in Figure 11(b). As seen in Figures 2(a) and (b), the polymer nanosphere multilayered organization of the aromatic polyamide is a double-particle layered structure. In other words, the period of two particles was confirmed by XRD, demonstrating that "upper" and "lower" regions of the particle in terms of electron density were present. That is to say, a hydrophilic group is localized to the bottom in the particle. On the other hand, the multiparticle layers formed by carbazole-based nanospheres were constructed from a single-particle layered structure. That is to say, there were no upper and lower regions of the particles.

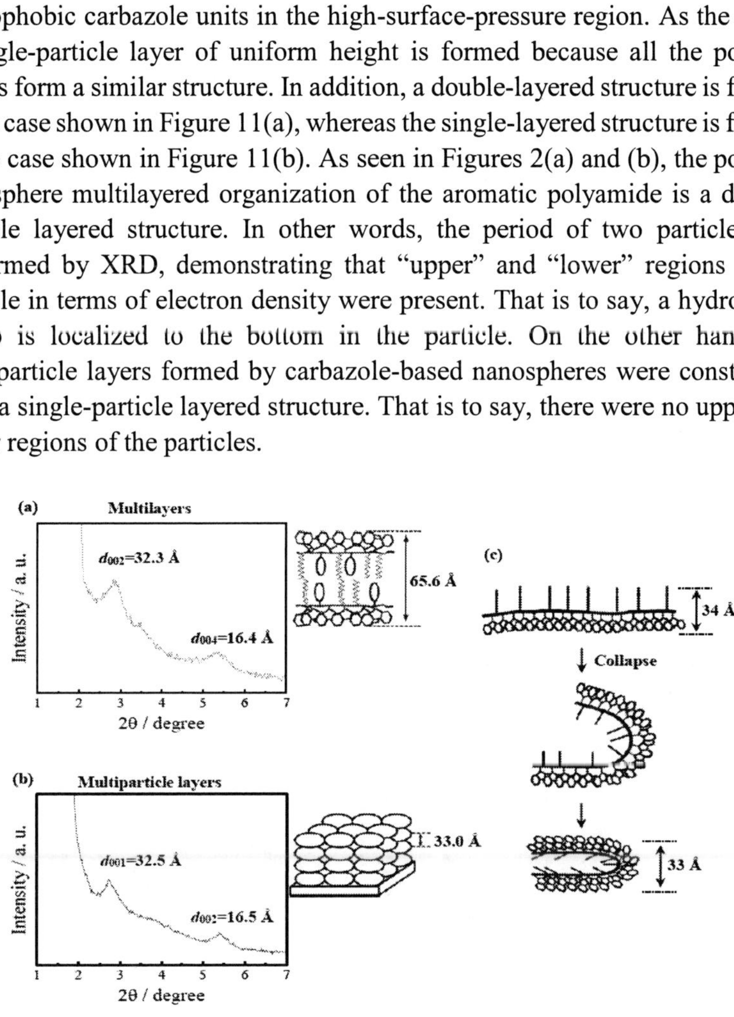

Figure 11. Out-of-plane XRD profiles and schematic illustrations of (a) LB multilayers transferred at 5 mNm^{-1} and (b) multiparticle layers transferred at 30 mNm^{-1} of NVCz:OA:FF_{10}EA = 4:1:1 copolymer; and (c) formation mechanism of single-particle layer on the water surface of ternary comb copolymers.

The formation of a layered organization using inorganic particles (as opposed to organic molecules) may also be attained in organo-modified inorganic compounds. Interestingly, it is possible to fabricate a single-particle layered structure and a double-particle layered structure simultaneously. Figure 12 shows a schematic illustration of the formation process of organo-modified ZrO_2. The organo-modifying agent covers the entire ZrO_2 particle surface, eliminating any distinction between upper and lower layers. Figure 13(a) shows the π–A isotherm of a single-particle layer of organo-modified ZrO_2 on the water surface. This single-particle layer forms a stable condensed film at the air/water interface. Figure 13(b) shows AFM images of a single-particle layer of organo-modified ZrO_2 at a different magnification. The obtained AFM images imply formation of a hierarchical structure of aggregated particle. The image at 4×4 μm^2 indicates that there are particles at about 200 nm diameter in the single-particle layer. Actually, it is found that these particles are aggregated and composed by fine ones at about 5–6 nm diameters by high resolution AFM image. The particle size of organo-modified ZrO_2 corresponds to a thickness of 5–6 nm in the single-particle layer. Figure 13(c) shows the out-of-plane XRD profile of multiparticle layers of organo-modified ZrO_2. The long period along the (00*l*) plane was 5.2 nm, which corresponds to a single-particle period. Figure 13(d) shows the structural color of multiparticle layers of organo-modified ZrO_2. As with the systems of polyamides and ternary comb copolymers with carbazole, the construction of a single-particle layer of uniform height, a highly ordered layered structure, and the appearance of structural color is attained in this inorganic system. In this case, however, only a single-particle layered structure is formed.

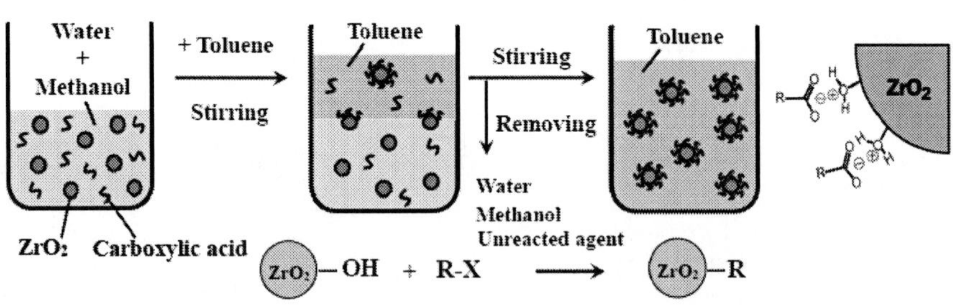

Figure 12. Schematic illustration of the formation process of organo-modified ZrO_2.

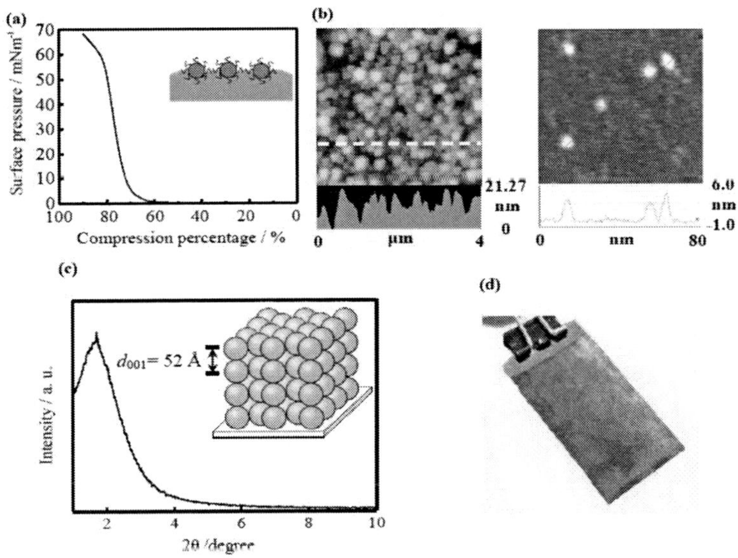

Figure 13. (a) Surface pressure–area isotherm of single-particle layer on the water surface of organo-modified ZrO_2. (b) AFM images of single-particle layer of organo-modified $ZrO_2$with different magnification. (c) Out-of-plane XRD profile of multiparticle layers of organo-modified ZrO_2, and (d) Appearance of structural color of multiparticle layers of organo-modified ZrO_2.

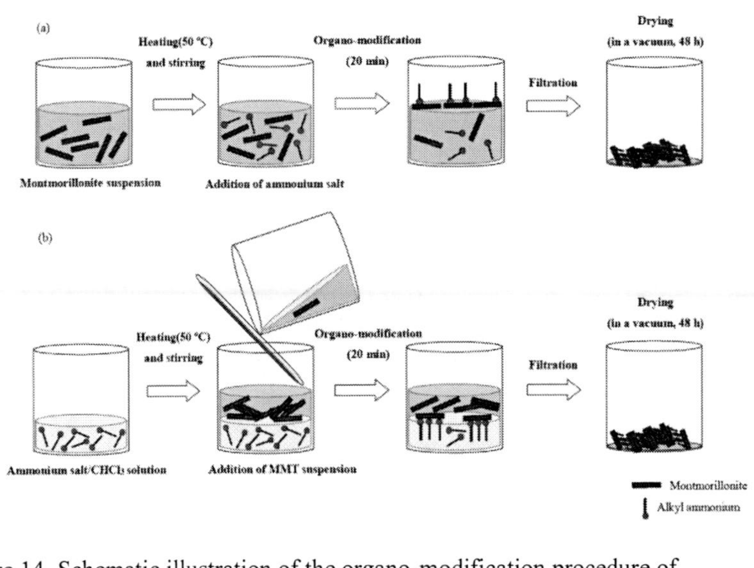

Figure 14. Schematic illustration of the organo-modification procedure of montmorillonite surface at chloroform/distilled water interface.

It is possible in certain cases to fabricate a double-particle layered structure in an inorganic system. We have previously reported a study on LB films of organo-modified clay [44, 45]. Figure 14 shows a schematic illustration of the organo-modification of a montmorillonite surface at a chloroform/distilled water interface. Figure 15(a) shows the π–A isotherm of an organo-modified montmorillonite particle monolayer on the water surface. This single-particle layer also forms a stable condensed film at an air/water interface. Figure 15(b) shows the AFM image of a Z-type monolayer of organo-modified montmorillonite particles. The surface coverage ratio of alkylated surface-treatment agent to montmorillonite indicated high coverage. Figure 15(c) shows the out-of-plane XRD profile of LB multilayers of organo-modified montmorillonite. A highly order layer structure with a period of approximately 60 nm was formed in this multiparticle layer. This long period corresponds to the double-particle period of organo-modified montmorillonite. Figure 15(d) shows a schematic illustration of this double-particle layered structure of organo-modified montmorillonite in the LB film. That is to say, a double-particle layered structure with head-to-head and tail-to-tail conformation is formed in this case.

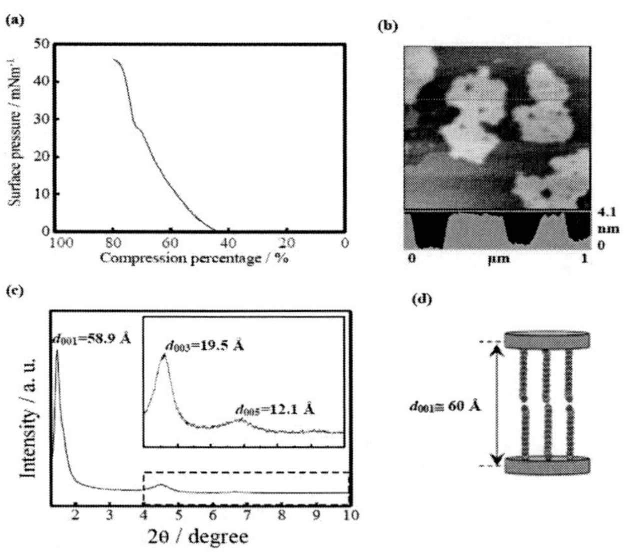

Figure 15. (a) Surface pressure–area isotherm of monolayer on the water surface of organo-modified montmorillonite particle; (b) AFM images of Z-type monolayer of organo-modified montmorillonite particle; (c) Out-of-plane XRD profile of LB multilayers of organo-modified montmorillonite; and (d) Schematic illustration of double-layered structure of organo-modified montmorillonite in the LB film.

1.7. Conclusion

A multiparticle layered organization of hydrophobic polymers was constructed by the LB technique. This organization was composed of built-up particle layers with highly regular arrangement along the *c*-axis. The particle arrangement was estimated by performing out-of-plane XRD analysis and AFM observation. From the XRD analysis, it was found that multiparticle layers of aromatic polyamides form a periodic structure of 5–6 nm along the c-axis. In addition, formation of a single-particle layer with a uniform height of 2.5–3.0 nm was confirmed by AFM. Because the long spacing estimated by XRD was almost twice the particle height, it was supposed that a double-particle layered structure was formed by applying the LB technique to amphiphilic particles at the air/water interface. However, in the case of a ternary comb copolymer system with carbazole units, only a single-layered structure with a 3.3 nm particle period was formed. Thus, it is possible simultaneously to fabricate a single-particle layered structure and a double-particle layered structure in the construction of multilayered polymer nanospheres for several hydrophobic polymer derivatives.

2. Control of Fine Structure in "Polymer Nanosphere Multilayered Organization" and Enhancement of Its Optical Property

2.1. Introduction

Non-covalent molecular organizations fabricated by the high control of molecular arrangement have demonstrated excellent functionality, which cannot be achieved in the case of single molecules, through the various intermolecular interactions occurring in that organization and cooperative phenomena based on that organization [1, 3, 4]. In addition, studies on the morphogenesis of molecular organization are beginning to elucidate the origin of life [55, 56] as well as induce function enhancement molecular devices [5]. Further, an extension of molecular functions may be possible by controlling molecular packing at the sub-nanometer level associated with mesoscopic morphology for achieving hierarchical arrangement [57, 58]. Of the several reported examples of "morphological control at the interface," the following ones are noteworthy;

viz., formation of nanofibers [59], nanowires [60, 61], nanocoils [62]. Nanoribbons [63], sea-island structures [64], rods [65], gyroids [66], lamellae [67], and honeycombs [68]. In recent years, "polymer nanosphere multilayered organization" has been proposed (Figure 16(a)) [26, 69]. In the previous studies, single particle layer has formed by using comb copolymers with reduced hydrophilicity than an amphiphilic polymer having an ability of formation of monomolecular film at air/water interface. Polymers with relative small amount of hydrophilic group form the interfacial particles in order to reduce the contact area with water at the air/water interface. Strictly speaking, the "collapsed film (multilayers)" units of dozens which stacked only orderly monomolecular layer thickness along the height direction, are aggregated. As a result, the height of the particles very homogenized. Multi-particle layers which is stacked one by one single particle layer is a colloidal crystal-like, and about uniform thickness to occur the structural color. This multi-particle layer organization is called as the "polymer nanosphere multilayered organization." In these studies, since polymer particles are layered regularly at approximately 5 nm periods, a clear layered period of particles and higher-order reflection are clearly observed by their out-of-plane XRD profile. The emission intensity of hydrophobic polymers containing a luminescent functional group is enhanced by the formation of multi-particle layers. This enhancement is attributed to the dense packing of light-emitting functional groups in the particles. Furthermore, in this section, an improvement in the layered regularity and the shift of the long period are mentioned for the new high-density, low-defect particle films. Here, a monolayer of the hydrophobic polymer is collapsed at the air/water interface, resulting in the formation of multilayers. Since the hydrophobic polymer avoids contact with the water surface, the collapsed film is stable (Figure 17). The "polymer nanosphere multilayered organization" is formed by aromatic polyamides, N-vinylcarbazole derivatives, polystyrene, and polyguanamine [69]. In the case of polymer monolayers, collapse of monolayer is sometimes referred to as a (three-dimensional) crystal transition [70]. Several tens of collapsed monolayer unit aggregate, resulting in the formation of nanoparticles at the air/water interface. Since the height of the particle corresponds to that of a collapsed monolayer, homogeneous height is observed only for a single bilayer or three molecular layers [24, 71]. On the other hand, since the particle size of the film plane depends on the aggregation number of the aggregated collapsed monolayer units, it is not necessarily uniform. However, clearly observed layered periods and high-order reflection are confirmed by XRD, which was employed to evaluate the height direction for multi-particle layers. This is similar to the formation of colloidal photonic crystals [72, 73] of single nanometer periods using organic molecules. As a result, this polymer nanosphere multilayered organization is expected to provide a wide range of applications, such as use in optical filters, stained glass, organic optical devices, water-repellent sheets, and reflective material (Figure 16(b)).

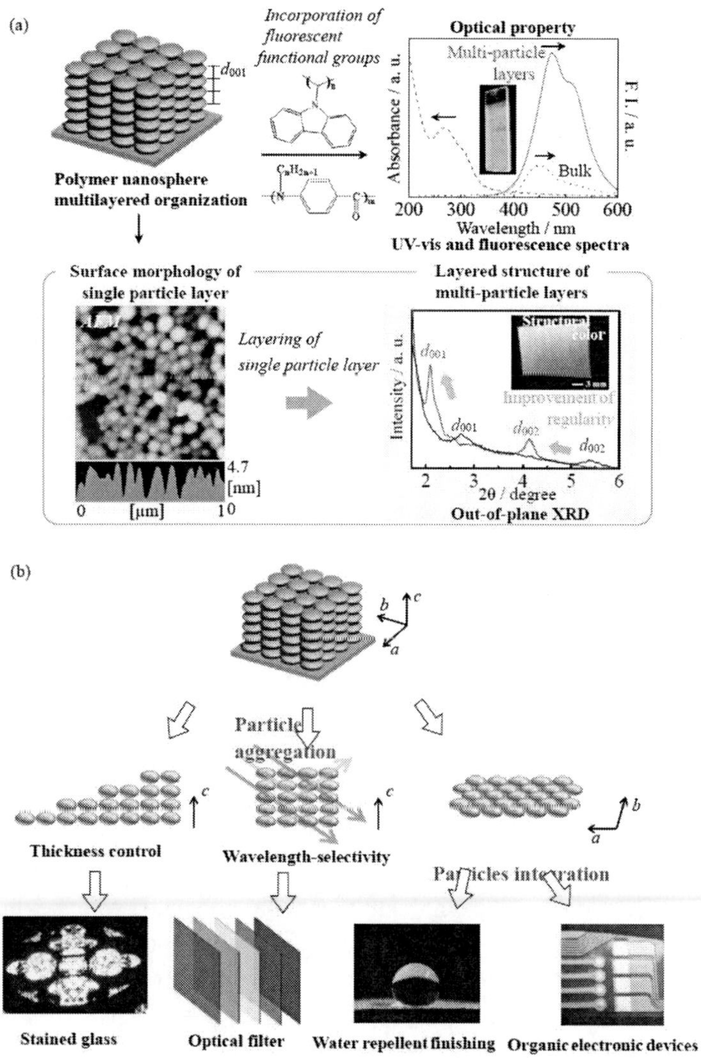

Figure 16. (a) Morphology, particle arrangement, and optical property of "polymer nanosphere multilayered organization." (b) Schematic of the potential applications of "polymer nanosphere multilayered organization."

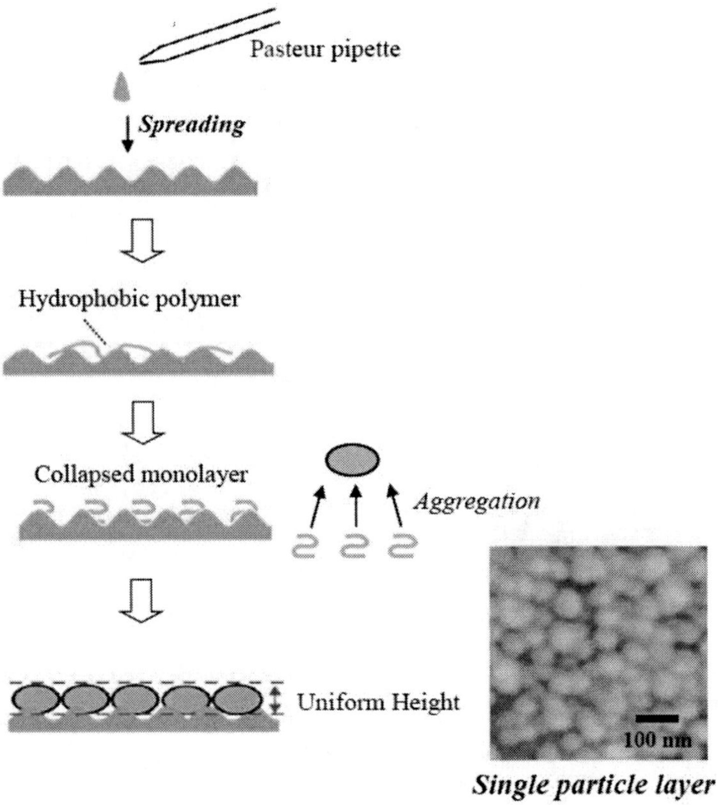

Figure 17. Schematic illustration of formation, structure, and function of polymer nano-sphere multi-layered organization.

Herein, functionalization *via* the nanostructure control of internal particles, that is, the elucidation of the relationship between the structure and function, was described. Optical properties were facilitated by the introduction of π-conjugated functional groups in the hydrophobic polymers, which exhibited an interesting change in the particle internal structure. A possible rearrangement occurs through the induction of enhancement of π–π interactions [74, 75] in the densely aggregated collapsed polymer monolayer. Dense π–π stacking interactions of the emitting moiety also led to the enhancement of its optical property.

Unfortunately, at this point, the detailed conditions under which nanoparticles are formed at the air/water interface are still unclear. The most likely factor for inducing their formation is the mobility [76] of the polymer on the water surface, without which the collapse of the polymer monolayer and formation of a single-particle film with subsequent aggregation cannot be achieved. The control of the particle size and the creation of a high-density, low-defect particle layer [77], which has not been achieved yet, can be achieved by a detailed investigation of conditions for particle formation. Compared to the size of the organized molecular film, that of the integrated particle layer containing constituent units is large, with relatively weak interaction between the structural units. Take advantage of these characteristics, high-quality particle-aggregated organization is formed.

In the second section of this capter, the formation of particles at the air/water interface was investigated by using a ternary comb copolymer having both hydrocarbon and fluorocarbon side chains. In addition, the formation of high-density and low-defect particle layers was attempted by repeated compression–relaxation [44]. Furthermore, the precise analysis of the internal particle nanostructure was performed to elucidate the relationship between the structure and function contributing to the enhanced light-emitting function. Herein, the formation of two types of side-chain crystals and arrangement of carbazole rings have been confirmed by thin-film XRD and deconvolution of fluorescence spectra [78].

2.2. Materials

Ternary comb copolymers used in this study were obtained by the copolymerization of NVCz with OA and $FF_{10}EA$ at various monomer ratios. $FF_{10}EA$ is a very good orientation as fluorinated monomer and is a solid state at room temperature. After the polymerization and forming a comb polymer (polymer brushes), it is possible to form side-chain crystals. In the case of FF_8EA having short length of fluorinated chain, this monomer is a liquid state at room temperature. Poly-FF_8EA side chain form the intermediate state of the gel and the side chain crystal. Therefore, it contains somewhat also amorphous area. Copolymerization was conducted in an acetone solution at 50°C for 48 h using azobisisobutyronitrile as the initiator. These monomers and the initiator were purchased from Tokyo Kasei Co. Ltd. and Daikin Fine Chemicals Co. Ltd., respectively, and were used without further purification. The compositions of the ternary comb copolymer were determined by ^{1}H NMR (EX270 NMR, Nihon Denshi Co. Ltd.) spectroscopy. The tacticity of the fluorinated homopolymer obtained by ^{1}H NMR analysis [79] was found to be almost syndiotactic (diad: 58%) according to the reference [80]. In this study, we synthesized ternary copolymers with molecular weights of approximately 4.43 × 10^4 (molecular weight distribution, $M_w/M_n \approx 2.14$) estimated by gel-permeation chromatography measurements.

According to the theory of the *Q–e* scheme proposed by Alfrey and Price [41], these ternary comb polymers form alternating copolymers of NVCz and long-chain acrylates. In this case, the *e* values of NVCz, OA, and $FF_{10}EA$ are −1.40, +1.12, and +0.66, respectively. From these values, it appears that the NVCz:long-chain acrylate copolymers form almost ideal alternating copolymers. Particularly, in this study, the NVCz:OA:$FF_{10}EA$ = 4:1:1 copolymer, which include high contents of hydrophobic carbazole along with an equal ratio of immiscible amphiphilic monomer units, was mainly used as starting materials for the formation of single-particle layers according to a previous report [69].

2.3. Procedures

2.3.1. Formation of Monolayers on the Water Surface and Estimation of Molecular Arrangement

The π–A isotherms were measured using a USI-3-22 film balance (USI Co. Ltd.) at 15°C. Monolayers and single-particle layers prepared from chloroform solutions (approximately 10^{-4} M) were formed on distilled water (resistivity is approximately 18 MΩ·cm). First, after the evaporation of chloroform from the solutions for 5 min, the π–A isotherms were recorded at compression speeds of 1.2, 4.8 (standard in this study), and 9.6 cm^2/min^{-1}. In these copolymer system, NVCz:OA:$FF_{10}EA$ = 2:1:1 copolymer form the monolayer with quite homogeneous, flat, and amorphous surface morphology by enhancement of π-π interaction between NVCz units [26]. In this case, phase separation between the hydrogenated and fluorinated side chains is impeded. On the other hand, hydrophobic polymer having few hydrophilic groups like a polyamide, polystyrene, etc, indicated single particle layer at air/water interface. Therefore, NVCz:OA:$FF_{10}EA$ = 4:1:1 and 6:1:1 copolymers with high contents of hydrophobic carbazole units had also formed the single particle layer at air/water interface and indicated transition from monolayer to single particle layer [69].

Next, monolayer films prepared by the LB method at 10, 15 (standard in this study), and 20°C were transferred onto glass for XRD studies and mica for AFM studies.

The surface morphologies of the transferred films were observed using a scanning probe microscope (Seiko Instrument, SPA300 with an SPI-3800 probe station) and microfabricated rectangular Si cantilevers with integrated pyramidal tips by applying a constant force of 1.4 Nm^{-1}.

The long spacing between the layer structures along the *c*-axis was estimated using an out-of-plane X-ray diffractometer (Rigaku, Rint-Ultima III, Cu*K*α radiation, 40 kV, 40 mA) equipped with a graphite monochromator. The in-plane spacing of the two-dimensional lattice of the films was determined by XRD (Bruker AXS, MXP-BX, Cu*K*α radiation, 40 kV, 40 mA) equipped with a parabolic graded multilayer mirror [46, 47].

UV–visible absorption spectra and fluorescence spectra of the LB films on quartz plates were recorded on a spectrophotometer (Jasco V-650) and a spectrofluorometer (Jasco FP-6500; λ_{ex} = 296 nm), respectively.

2.3.2. Creation of High-density, Low-defect Single-layer Films of Polymer Nanospheres

This study investigated the formation of single-particle layers of "polymer nanosphere multilayered organization," high-density, low-defect particle layers, as well as the formation conditions of single-particle layers at the air/water interface, precise analysis of the internal fine structure of particles, and enhancement of optical property.

Figure 18 shows the formation of high-density, low-defect single-particle layers of the NVCz:OA:FF_{10}EA = 4:1:1 copolymer. The evidence for their formation was supported by AFM, out of plane XRD, and a change in the fine crystallite size based on the Scherrer equation (the Scherrer constant K = 0.94 was used). In a previous study [69], single-particle layers of the polymer nanospheres formed at the air/water interface have been observed to exhibit many defects and low density (Figure 18(a)). In contrast, the film (Figure 18(b)) obtained in this study exhibits low defects and high particle density. Although it is exaggerated perspective, this low-defect particle layer indicates that the packing of the system resembles hexagonal packing at a plurality of positions. (The corresponding part has described by the auxiliary line.) Although the out-of-plane XRD profile of the multi-particle layers, as shown in Figure 18(a), exhibits clear reflections (Figure 18(c)), the differences are obvious by comparison with the high-density film (Figure 18(d)). Diffraction peaks clearly become sharp. The third order reflection peak also appeared although this intensity was quite weak. (Other information has been confirmed from influence on the underlayer.) The difference at approximately 200 Å is approximately calculated by the comparison of the diameter of the crystallite size in a direction perpendicular to the (001) plane by using the Scherrer equation [81]. Notably, the peak value shifts to low angles. Although a layered organization of particles is composed of exactly the same compound and prepared under the same formation conditions, there is an approximately 10 Å difference in the long period value (as compared to the values of $d_{002} \times 2$). It is indicated that the distance between the centers of gravity in previous report decreased by the "subduction" of the particles shown in the illustrations (Figure 18(c)). In the case of high-density, low-defect particle layers, the particles in the upper layers are ordered on top of the particles in the lower layers.

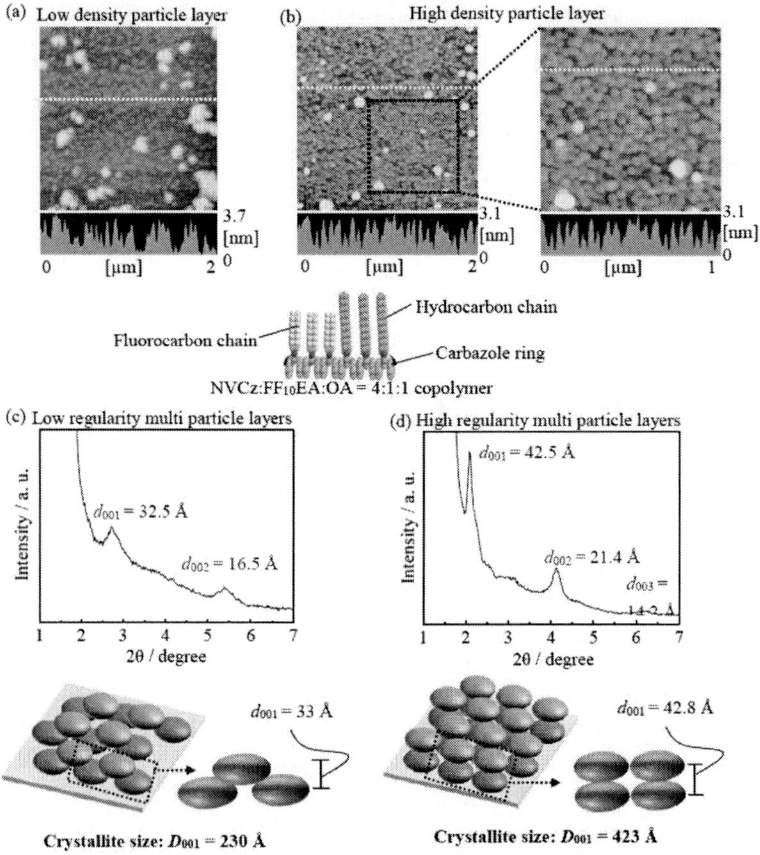

Figure 18. AFM topology images of Z-type single-particle layers of the NVCz:FF_{10}EA:OA = 4:1:1 copolymer on a solid substrate (a) before and (b) after "the alternate compression–relaxation method." Out-of-plane XRD profiles of the 20 layers of the NVCz:FF_{10}EA:OA = 4:1:1 copolymer obtained by the LB method (c) before and (d) after "the alternate compression–relaxation method."

Next, the method of forming this high-density particle layer and the reversibility of particle formation are described. Figure 19 shows the reversibility of the π–A curve at 15°C. At the initial compression stage, a two-dimensional phase transition is observed, i.e., monolayer–single-particle layer transition, which corresponds to the collapse of the monolayer. By hysteresis loop measurements at film expansion after particle film formation, an irreversible π–A curve is obtained. In other words, the single-particle layer of polymer nanospheres that was once formed at the air/water interface does not reverse back to the monolayer. In addition, during hysteresis loop measurements, the value of the limiting area along the horizontal axis is gradually narrowed. Actually, when this process is performed six times, the 5th and 6th curves of the hysteresis loop almost coincide. As a result, the particle layer attains the highest low-defect, high-density state. Hence, we refer to this process as the "alternate compression–relaxation method" for generating a single-particle layer on the water surface. For a system in which the size of components is large and the interaction between the constituent units is relatively weak, a low-defect layer can be effectively formed by this process.

Figure 20 provides morphological evidence for the above speculation and internal state of the particles indicated from size information. Figure 20(a) shows AFM topology images of the monolayer state before phase transition. Since a phase transition is observed at 10 mNm^{-1}, a monolayer is clearly formed during transfer at 5 mNm^{-1}. At the transfer after the phase transition of the first compression process (35 mNm^{-1}), a low-density single-particle layer with many defects is observed (Figure 20(b)). Non-existent part of the particles is corresponded to the exposed-substrate. After alternate compression–relaxation, there is an obvious improvement in the particle density (Figure 20(c)). Judging from the values of the limiting area observed in the π–A isotherm and the height data obtained from AFM, the polymer nanospheres are composed of a collapsed unit of polymer monolayer-like bilayer as their structural component (Figure 20(d)). The average diameter of the nanospheres is 53.7 nm, and their particle distribution is shown in Figure 20(e). In a simple calculation, a collapsed film of approximately 60 units is assembled with an aspect ratio of 1:10. That is to say, the actual interfacial nanoparticles are ellipsoid.

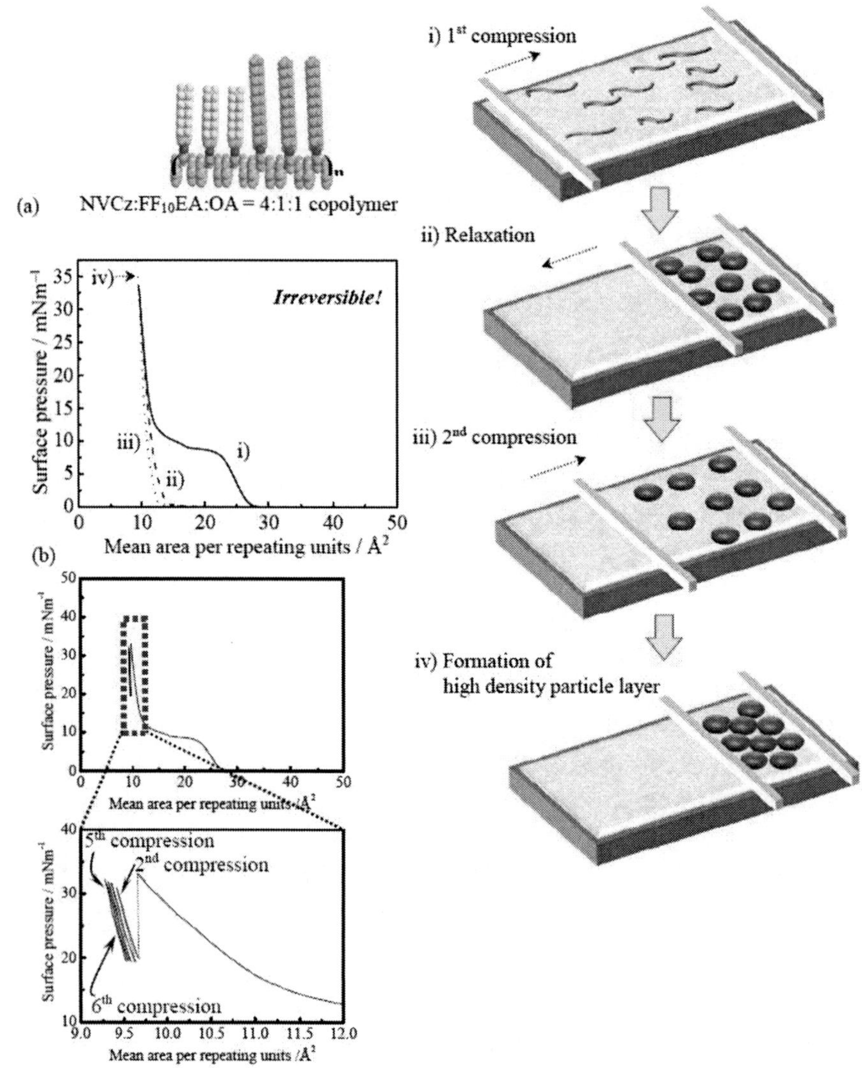

Figure 19. (a) Hysteresis loop of π–A isotherms at 15°C of the monolayer of the NVCz:FF_{10}EA :OA = 4:1:1 copolymer (solid line: 1st compression, dashed line: relaxation, dotted line: 2nd compression), and schematic of “the alternate compression–relaxation method.” (b) Changes in limiting area by compression times of π–A isotherms at 15°C of the monolayer of the NVCz:FF_{10}EA :OA = 4:1:1 copolymer.

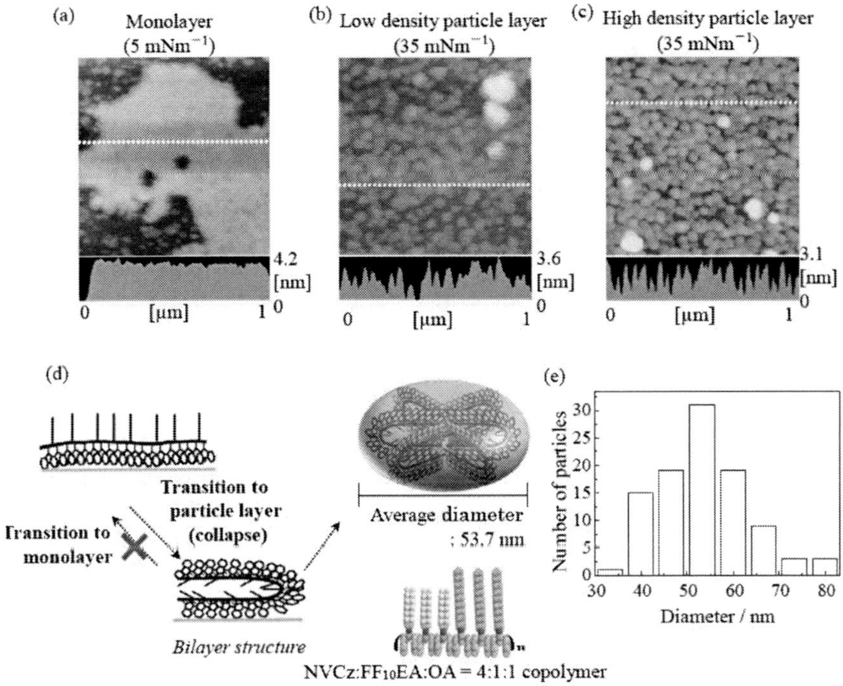

Figure 20. AFM topology images of the Z-type monolayer or single-particle layers of the NVCz:FF_{10}EA:OA = 4:1:1 copolymer on a solid substrate transferred at (a) 5, (b) 35 mNm^{-1}, and (c) 35 mNm^{-1} by "the alternate compression–relaxation method." (d) Schematic of the formation of single-particle layers on the water surface; (e) Size distribution of the single-particle layer.

Currently, the particle film of higher density and further low defects have been studied. Now, the missing point of view is the uniformity of the particle diameter. If there is a large distribution in particle diameter, it essentially occurs defects in density integrated particle layers. The homogenization of the particulate diameter is determined by uniformity of the chain length of the chemical structure of polymers. In other words, this is a narrow molecular weight distribution of the polymer. Since the collapsed monolayer were regularly folded main chain, it was homogeneous height of the particles. The particle size in the two-dimensional plane is important key to the homogenization of the molecular chain length of the polymer as a component of a particle. Therefore, the fusion of the polymer synthesis and film-forming technologies is considered to be directing the possibility of widening the application of the particle film material in the future.

2.4. Study of Formation Conditions of Single-Particle Layers at the Air/Water Interface

Figure 21 shows the dependence of compression speed and subphase temperature on the formation condition of the interfacial particle layer. As indicated by the π–A curves in Figure 21(a), the transition from the single-particle layer to the monolayer does not appear at low-speed compression. Moreover, at a slow compression speed of 1.2 cm^2/min, a two-dimensional transition is not observed, and after the transfer at 35 mNm^{-1}, only a homogenous monolayer is observed (Figure 21(b)). Even if a compression speed of 9.6 cm^2/min is applied, a large difference in particle size distribution is not observed (Figures 21(c) and (d)).

Figure 21(e) shows the temperature dependence of the π–A isotherms of the NVCz:FF_{10}EA:OA = 4:1:1 copolymer [69]. As described above, the phase transition from a monolayer to a single-particle layer is confirmed at a subphase temperature of 15°C and at around 10 mNm^{-1}. However, at subphase temperatures of 10 and 20°C, such a phase transition is not observed in both isotherms. However, the limiting areas of both isotherms are quite different. The limiting area of the π–A curve at 10°C is approximately equal to that of the monolayer state at 15°C. On the other hand, at a subphase temperature of 20°C, the limiting area is significantly reduced. From the AFM images of the transferred film on the solid substrate, morphologies of the monolayer and single-particle layers were observed at 10 and 20°C (Figures 21(f) and (g), respectively). Also, as compared to the subphase temperature of 15°C, at 20°C, there is no significant difference in the particle size distribution (Figure 21(h)).

Figure 22 shows a schematic phase diagram of the NVCz:FF_{10}EA:OA copolymers between monolayer and single-particle layers by employing surface pressure, subphase temperature, and compression speed as parameters. When the ratio of NVCz to the amphiphilic side-chain unit is low, a very homogeneous monolayer is created. Especially, for the NVCz:FF_{10}EA: OA=2:1:1 copolymer, a new type of a polymer nanosheet with a homogeneous amorphous surface is formed according to a previous study [40]. This observation matches our finding for the NVCz:FF_{10}EA:OA = 4:1:1 copolymer, as well as 3:1:1 and 6:1:1 copolymers [69], in that all of the copolymers exhibit the same phase transition behavior. On the other hand, for the NVCz : FF_{10}EA : OA = 1 : 0 : 0 copolymer, i.e., a poly-NVCz homopolymer, only a single-particle layer state is formed under all conditions. Under the conditions of high temperature, high pressure, and high compression speed, the transition from the copolymer monolayer to a single-particle layer occurs. In other words, the high mobility of the polymer chains at the air/water interface is found to be suitable.

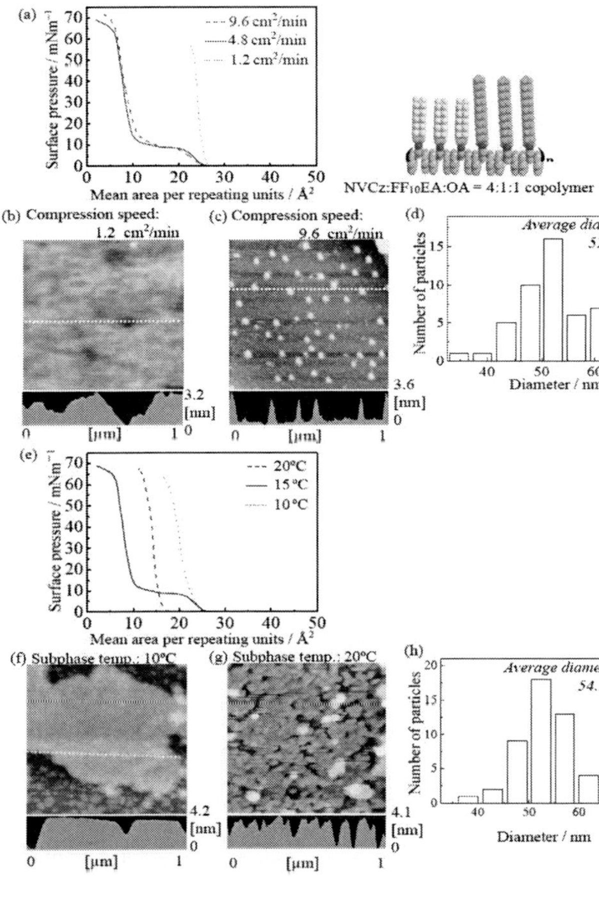

Figure 21. (a) π–A isotherms of the monolayer at the air/water interface at different surface compression speeds (15°C, compression speeds: 4.8 cm^2/min; solid line, 9.6 cm^2/min; dashed line, and 1.2 cm^2/min; dotted line), and AFM images of the Z-type monolayer and single-particle layer on a solid transferred at 35 mNm^{-1} with compression speeds of (b) 1.2 min/cm^2 and (c) 9.6 min/cm^2, respectively. (d) Size distribution of the particles in Figure 21(c). (e) π–A isotherms of the monolayer on the water surface at different subphase temperatures (compression speed: 4.8 cm^2/min; 10°C; dotted line, 15°C; solid line, and 20°C; dashed line), and AFM images of the Z-type monolayer and single-particle layer on a solid transferred at 35 mNm^{-1} with subphase temperatures of (f) 10°C and (g) 20°C. (h) Size distribution of the particles in Figure 21(g).

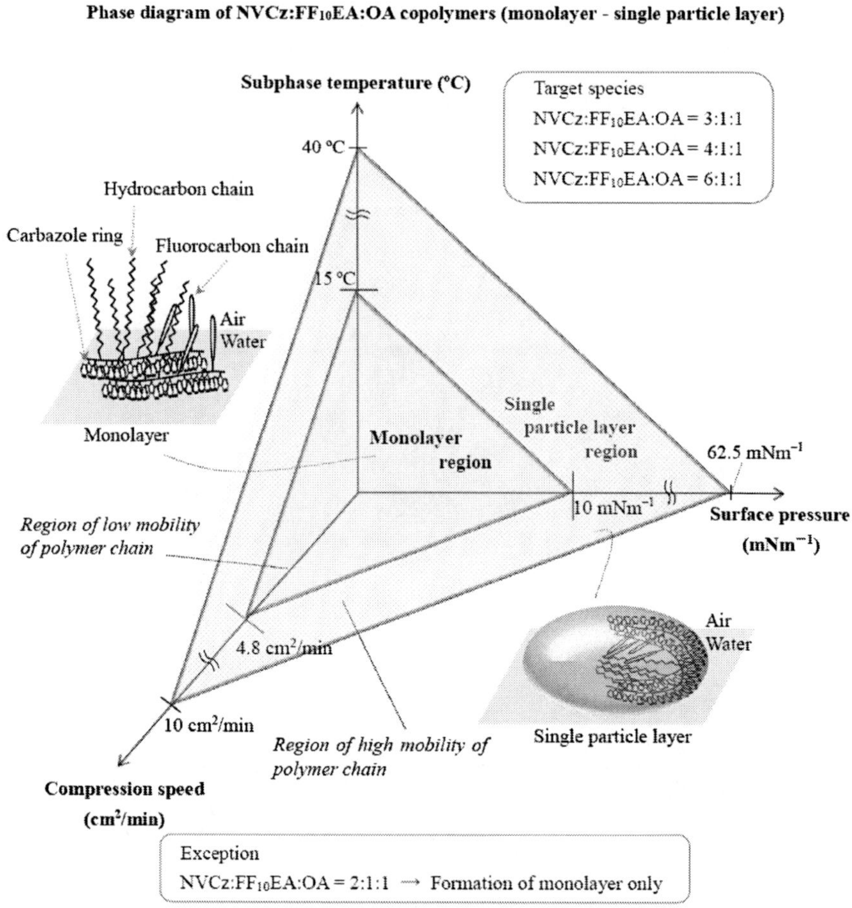

Figure 22. Phase diagram of the NVCz:FF_{10}EA:OA = 3:1:1, 4:1:1, and 6:1:1 copolymers at the air/water interface.

2.5. Fine Structure Analysis of Internal Particles

For estimating the internal fine structure of the polymer nanospheres, polarized UV–vis and IR spectra of the multi-particle layers of the NVCz:FF_{10}EA:OA = 4:1:1 copolymer were recorded; the results are shown in Figure 23. The transition moment of the π-conjugated system, observed by the polarized UV spectrum, and C = C double bonds of the carbazole ring, as examined by the polarization IR spectrum, typically show clear incident angle dependence, and the orientation angle of the functional groups calculated from their dichroic ratio is 32°. In other words, functional carbazole units in the polymer nanospheres have a regular oriented structure at certain angles.

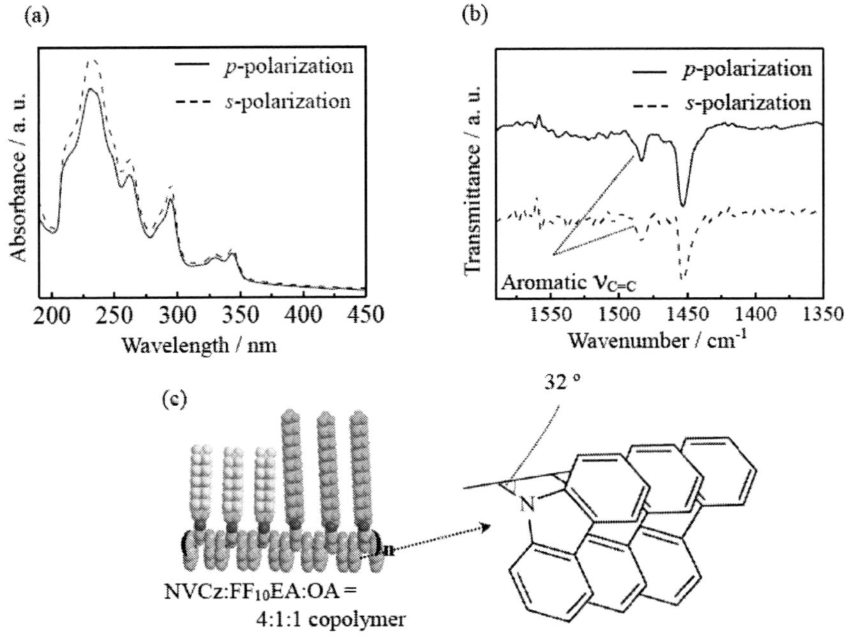

Figure 23. (a) Polarized UV–vis spectra and (b) polarized IR spectra of the LB multi-particle layers (15°C, 35 mNm^{-1}, and 4.8 cm^2/min). (c) Schematic of the orientation of the carbazole ring in the polymer nanospheres.

For estimating the crystallinity and packing system of fluorinated and hydrogenated side chains in the polymer nanospheres, in-plane XRD profile of the multi-particle layers of the NVCz:FF$_{10}$EA:OA = 4:1:1 copolymer were recorded; the results are shown in Figure 24. Two types of (100) reflection are observed, which correspond to the hexagonal close packing of each side chain. The left signal (d^{F}_{100} = 5.0 Å) is caused by the short spacing of fluorocarbon chains, and the right signal (d^{H}_{100} = 4.2 Å) is caused by one of the hydrocarbon chains. Therefore, it seems that both side chains form the phase-separated side-chain crystals (sub-cell [82]) inside of nanoparticles each other. Figure 24(b) also shows an illustration of sub-cell models of hydrocarbon and fluorocarbon chains. The formation of the side-chain crystals in the interlayer inside the nanoparticles is expected to be a stabilization factor for the bilayer structure with a folded main chain.

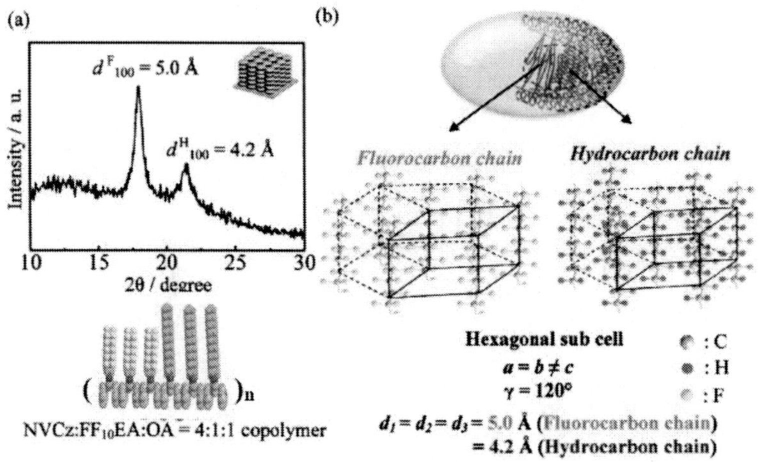

Figure 24. (a) In-plane XRD profile of the LB multi-particle layers of the NVCz:OA:$FF_{10}EA$ = 4:1:1 copolymer (15°C, 35 mNm^{-1}, and 4.8 cm^2/min). (b) Schematic models of two-dimensional side-chain crystals (sub-cell) of hydrocarbons and fluorocarbons.

2.6. Enhancement of Fluorescence Emission Properties

Figure 25 shows the comparison of the normalized intensity of the fluorescence spectra of multilayers and multi-particle layers of the NVCz:$FF_{10}EA$:OA = 4:1:1 copolymer with the same number of layers. This copolymer emits fluorescence at 370–450 nm. Upon photoexcitation, poly-NVCz exhibits a broad fluorescence spectrum in the violet–blue region, which is attributed to the overlap of emissions from different excimers. The high-energy emission centered at around 370 nm is ascribed to a partially overlapped structure involving only one eclipsed aromatic ring from each carbazole group, while the low-energy emission centered at around 420 nm is proposed to correspond to a "sandwich"-like conformation in which two carbazolyl groups attain an eclipsed or fully overlapped arrangement [63, 83]. Notably, the normalized emission intensity value of the multi-particle layer is approximately two times that of the multilayers. The carbazole ring exhibiting fluorescence with an oriented structure, as discussed in the preceding paragraph, is highly ordered and arranged. Suitable unit distance to the light emission of the carbazole excimer [84] is formed in a wide range, and it is considered to efficiently form the excited state.

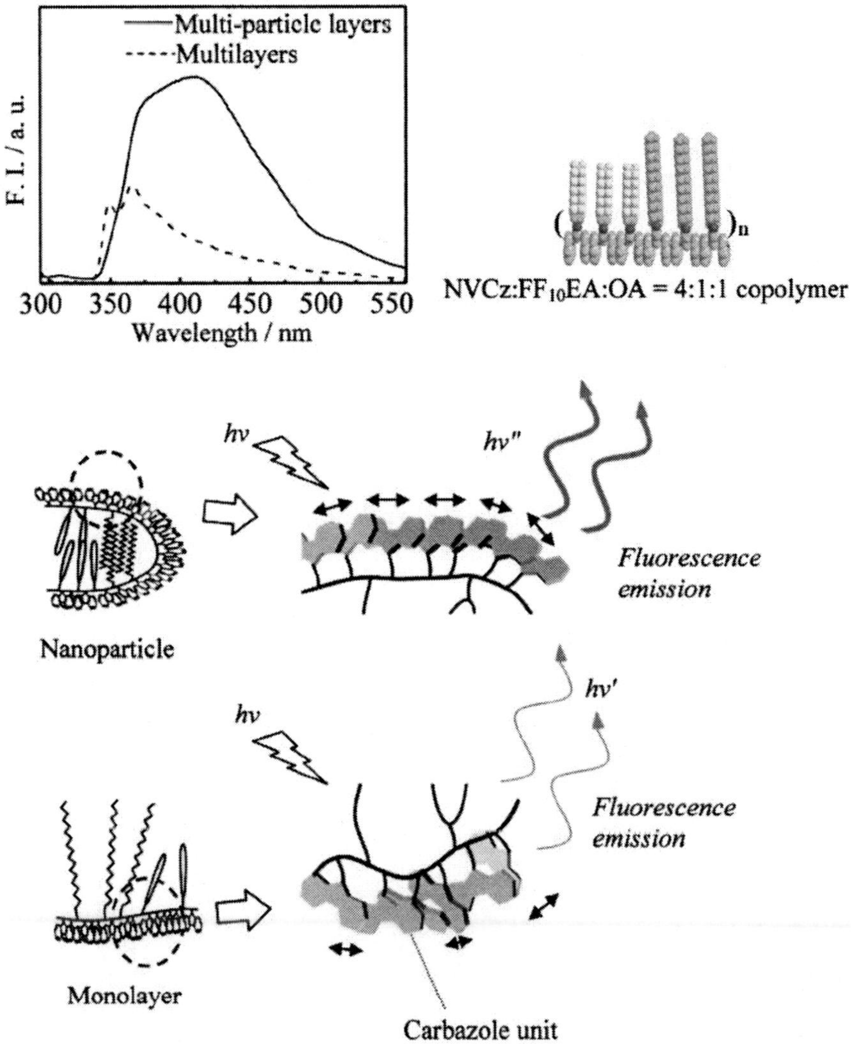

Figure 25. Fluorescence spectra of the LB multi-particle layers and multilayers of the NVCz:OA:$FF_{10}EA$ = 4:1:1 copolymer (15°C, 35 mNm^{-1}, and 4.8 cm^2/min).

This emission is also confirmed in dilute solutions of the NVCz:FF_{10}EA:OA = 4:1:1 copolymer (Figure 26(a)). Excimer emission is observed in dilute solutions, mainly originating from the intensity of the emission band, attributed to the partial overlap of the carbazole ring. Of course, in multi-particle layers, light emission by the carbazole excimer is also observed (Figure 26(b)). However, the intensity of the emission band is stronger, caused by the excimer emission based on overlap of the carbazole ring. From this result, it is considered that the carbazole rings arranged on the outermost layer inside the particles mainly undergo “sandwich”-like stacking. Compared with the multilayers (Figure 26(c)), the multi-particle layers exhibit an increase in the ratio of the partial (tilted) stack, which can be observed as a sharp emission band. However, since the ratio of the parallel stacks is not improved, predominant formation of this orientation state is characteristic of the “polymer nanosphere multilayered organization.”

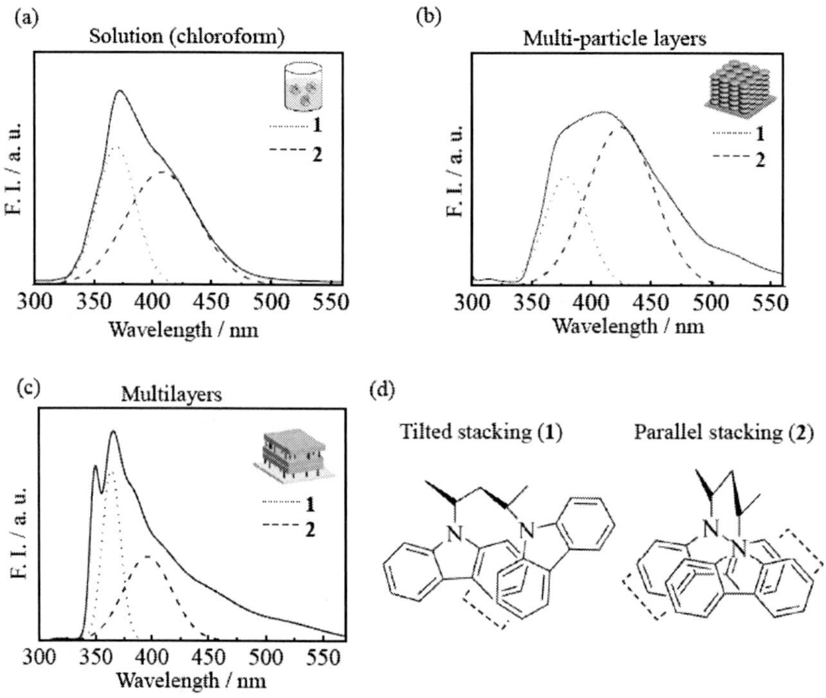

Figure 26. Fluorescence spectra of (a) chloroform solution (10^{-4} mol/L), (b) the LB multi-particle layers, and (c) the LB multilayers of the NVCz:OA:FF_{10}EA = 4:1:1 copolymer. (d) Schematic of the stacking of the carbazole rings in the polymer nanospheres.

Figure 27 shows an illustration of the hierarchical structure of "polymer nanosphere multilayered organization" of the NVCz:FF_{10}EA:OA ternary comb copolymer based on fine structure analysis and light-emitting function measurements. Figures 22, 24, and 27 describe the structural model and the oval shape (i.e., particles with large aspect ratio) to a large extent depending on the AFM observations. In the near future, the additional measurement with a transmission electron microscope needs to be verified. However, authors have certain confidence in ourselves to structure estimated by XRD, AFM, and π-A isotherm. It is an inference that twice value of monolayer thickness corresponds to particle height and interlayer side chain crystal supports this structure since monolayer are collapsed and polymer chain overlaps at height direction. A beautiful hierarchical structure is formed from the functional group level to the layered organization level. Such sophisticated molecular techniques with bottom-up technology would introduce avenues for new science and for the development of new molecular devices [85].

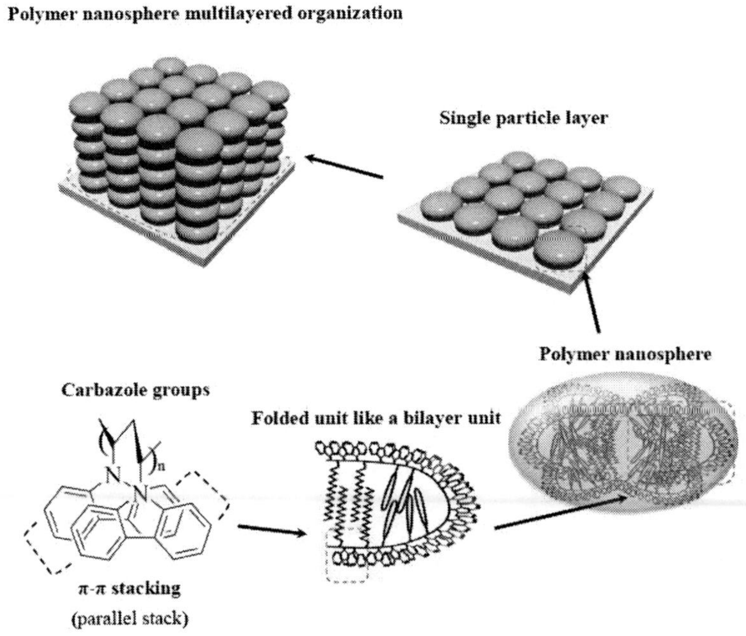

Figure 27. Schematic models of the hierarchical structure of "polymer nanosphere multilayered organization" of the NVCz:OA:FF_{10}EA = 4:1:1 ternary comb copolymer.

2.7. Conclusion

In this study, fine structure analysis and new functionality produced by "polymer nanosphere multilayered organization," a new type of molecular organization, are described, as well as the relationship between molecular orientation and functionality. The polymer nanosphere multilayered organization is an elaborate structural material formed by the uniform layering of polymer single-particle layers on the water surface, which have uniform height along the *c*-axis. In this paper, 1. formation of high-density and low-defect single particle layer, 2. precise analysis of hierarchical structure from the internal particle, and 3. enhancement of optical properties of this organization have been studied. Applying the "alternate compression-relaxation process," high-density, low-defect particle layers are easily formed, with a clear increase in their crystallite sizes. In this process, it was also carried out scrutiny of the particle formation conditions. Under the conditions of high subphase temperature, high pressure, and high compression speed, relatively easy formation of nanoparticles of the hydrophobic polymer occurs. The "Polymer nanosphere multilayered organization" is found to be a hierarchical organized structure. In the case of a ternary copolymer containing a carbazole ring, particle diameter is the size of a few tens of nm, though the particle height is the thickness of about 5 nm. One particle is formed by the assembly of approximately 60 units of a collapsed monolayer-like double layer. Therefore, particle height is extremely homogeneous. This structure is stabilized by the formation of side-chain crystals in the interlayer, with oriented π-π stacking of carbazole rings. As a result, fluorescence emission intensity is enhanced, and spectral changes are observed by a change in ring stacking. In other words, reasonable structure / function relationships have expressed. These results support the formation of a beautiful hierarchical structure from the functional group level to the layered organization level.

REFERENCES

[1] Gaines, G. L., Jr. *Insoluble Monolayers at Liquid Gas Interfaces*; Wiley: New York, 1966.

[2] Roberts, G. G. *Langmuir-Blodgett Films*; Plenum Press: London, 1990.

[3] Petty, M. C. *Langmuir-Blodgett Films*; Cambridge Univ. Press: New York, 1996.

[4] Ulman, A. *An Introduction to Ultrathin Organic Films: From Langmuir-Blodgett to Self-assembly*; Academic Press: Boston, 1991.

[5] Kuhn, H.; Möbius, D.; Bücher, H. *Spectroscopy of Monolayer Assemblies, in: Physical Methods of Chemistry*; Weissberger, A., Rossiter, B. W., Eds.; Wiley: New York, 1972; Vol. 1, Part IIIB.

[6] Prusiner, S. B. Prions. *Proc. National Acad. Sci. USA* 1998, *95*, 13363-13383.

[7] Rost, B.; Sander, C. Prediction of Protein Secondary Structure at Better Than 70% Accuracy. *J. Mol. Biol.* 1993, *232*, 584-599.

[8] Sussman, J. L.; Harel, M.; Frolow, F.; Oefner, C.; Goldman, A.; Toker, L.; Silman, I. Atomic Structure of Structure from Torpedo Californica: a Prototypic Acetylcholine-binding Protein. *Science* 1991, *253*, 872-879.

[9] Keller, A. A Note on Single Single in Polymers: Evidence for a Folded Chain Configuration. P. Mag. 1957, *2*, 1171-1175.

[10] Till, P. H. The Growth of Single Crystals of Linear Polyethylene. *J. Polym. Sci.* 1957, *24*, 301-306.

[11] Fischer, E. W. Notizen: Stufen- Und spiralförmiges Kristallwachstum bei Hochpolymeren. Z. Naturforsch. 1957, 12, 753-754.

[12] Alexandre, M.; Dubois, P. Polymer-Layered Silicate Nanocomposites: Preparation, Properties and Uses of a New Class of Materials. *Mater. Sci. Eng. R-rep.* 2000, *28*, 1-63.

[13] Stankovich, S.; Dikin, D. A.; Dommett, G. H. B.; Kohlhaas, K. M.; Zimney, E. J.; Stach, E. A.; Piner, R. D.; Nguyen, S. T.; Ruoff, R. S. Graphene-Based Composite Materials. *Nature* 2006, *442*, 282-286.

[14] LeBaron, P. C.; Wang, Z.; Pinnavaia, T. J. Polymer-Layered Silicate Nanocomposites: an Overview. *Appl. Clay Sci.* 1999, *15*, 11-29.

[15] Kresge, C. T.; Leonowicz, M. E.; Roth, W. J.; Vartuli, J. C.; Beck, J. S. Ordered Mesoporous Molecular Sieves Synthesized by a Liquid-Crystal Template Mechanism. *Nature* 1992, *359*, 710-712.

[16] Castro, N. A. H.; Guinea, F.; Peres, N. M. R.; Novoselov, K. S.; Geim, A. K. The Electronic Properties of Graphene. *Rev. Modern Phy.* 2009, *81*, 109-162.

[17] Tarascon, J. M.; Armand, M. Issues and challenges facing rechargeable lithium batteries. *Nature* 2001, *414*, 359-367.

[18] Reimers, J. N.; Dahn, J. R. Electrochemical and In Situ X-Ray Diffraction Studies of Lithium Intercalation in Li_xCoO_2. *J. Electrochem. Soc.* 1992, *139*, 2091-2097.

[19] Tarascon, J. M.; Mckinnon, W. R.; Coowar, F.; Bowmer, T. N.; Amatucci, G.; Guyomard, D. Synthesis Conditions and Oxygen Stoichiometry Effects on Li Insertion into the Spinel $LiMn_2O_4$. *J. Electrochem. Soc.* 1994, *141*, 1421-1431.

[20] Amatucci, G. G.; Tarascon, J. M.; Klein, L. C. CoO_2, The End Member of the LixCoO2 Solid Solution. *J. Electrochem. Soc.* 1996, *143*, 1114-1123.

[21] Peumans, P.; Yakimov, A.; Forrest, S. R. Small Molecular weight Organic Thin-film Thinand Solar Cells. *J. Appl. Phy.* 2003, *93*, 3693-3723.

[22] Granstrom, M.; Petritsch, K.; Arias, A. C.; Lux, A.; Andersson, M. R.; Friend, R. H. Laminated Fabrication of Polymeric Photovoltaic Diodes. *Nature* 1998, *395*, 257-260.

[23] Nitzan, A. Electron Transmission Through Molecules and Molecular Interfaces. *Annu. Rev. Phy. Chem.* 2001, *52*, 681-750.

[24] Vaknin, D.; Bu, W.; Satija, S. K.; Travesset, A. Ordering by Collapse: Formation of Bilayer and Trilayer Crystals by Folding Langmuir Monolayers. *Langmuir* 2007, *23*, 1888-1897.

[25] Yan, H.; Sokolov, S.; Lytle, J. C.; Stein, A.; Zhang, F.; Smyrl, W. H. Colloidal-Crystal-Templated Synthesis of Ordered Macroporous Electrode Materials for Lithium Secondary Batteries. *J. Electrochem. Soc.* 2003, *150*, A1102-A1107.

[26] Fujimori, A.; Chiba, S.; Sato, N.; Shibasaki, Y.; Abe, Y. Surface Morphological Changes in Monolayers of Aromatic Polyamides Containing Various N-Alkyl Side Chains. *J. Phys. Chem. B* 2010, *114*, 1822-1835.

[27] Shibasaki, Y.; Abe, Y.; Sato, N.; Fujimori, A.; Oishi, Y. Direct Condensation Polymerization of N-Alkylated p-Aminobenzoic Acid and Packing of Rigid-Rod Main Chains with Flexible Side Chains. *Polym. J.* 2010, *42*, 72-80.

[28] Harada, T.; Matsuoka, H.; Ikeda, T.; Yamaoka, H. An Exact Evaluation of Salt Concentration Dependence of Interparticle Distance in Colloidal Crystals by Ultra-Small-Angle X-Ray Scattering. 3. Confirmation of Solid-Liquid Transition by Three-Dimensional Paracrystal Analysis. *Langmuir* 1999, *15*, 573-577.

[29] Jawalkar, S. S.; Adoor, S. G.; Sairam, M.; Nadagouda, M. N.; Aminabhaviin, T. M. Molecular Modeling on the Binary Blend Compatibility of Poly(vinyl alcohol) and Poly(methyl methacrylate): An Atomistic Simulation and Thermodynamic Approach. *J. Phys. Chem. B* 2005, *109*, 15611-15620.

[30] Prathab, B.; Aminabhavi, T. M.; Parthasarathi, R.; Manikandan, P.; Subramanian, V. Molecular Modeling and Atomistic Simulation Strategies to Determine Surface Properties of Perfluorinated Homopolymers and Their Random Copolymers. *Polymer* 2006, *47*, 6914-6924.

[31] Prathab, B.; Aminabhavi, T. M. Atomistic Simulations to Compute Surface Properties of Poly (N-Vinyl-2-pyrrolidone)(PVP) and Blends of PVP/Chitosan. *Langmuir* 2007, *23*, 5439-5444.

[32] Jawalkar, S. S.; Raju, K. V. S. N.; Halligudi, S. B.; Sairam, M.; Aminabhaviin, T. M. Molecular Modeling Simulations to Predict Compatibility of Poly (vinyl alcohol) and Chitosan Blends: a Comparison with Experiments. *J. Phys. Chem. B* 2007, *111*, 2431-2439.

[33] Riess, G.; Hurtrez, G. Bahadur, P. in: *Encyclopedia of Polymer Science and Engineering*; Mark, H. F., Kroscwitz, J. I., Eds.; Wiley: New York, 1985.

[34] Hamley, I. W. *The Physics of Block Copolymers*; Oxford University Press: New York, 1998.

[35] Watanabe, J.; Sasaki, S.; Uematsu, I. Structure and Mechanical Properties of Poly(γ-methyl d-glutamate) Films. I. Influence of the Casting Solvents. *Polym. J.* 1977, *9*, 451-458.

[36] Watanabe, J.; Nakata, M.; Watanabe, M.; Uematsu, I. Structure and Mechanical Properties of Poly(γ-methyl glutamate) Films. II. *Polym. J.* 1978, *10*, 569-582.

[37] Watanabe, J.; Sasaki, S.; Uematsu, I. Cholesteric Twisted Structure in Solid Films of Poly(γ-methyl-d-glutamate). *Polym. J.* 1977, *9*, 337-340.

[38] Watanabe, J.; Imai, K.; Gehani, R.; Uematsu, I. Structural Differences Between Two Crystal Modifications of Poly(γ-benzyl L-glutamate). *J. Polym. Sci., Part B: Polym. Phys.* 1981, *19*, 653-665.

[39] Watanabe, J.; Imai, K.; Gehani, R.; Uematsu, I. A Structure with an Irregular Layer Lattice in Poly(γ-methyl L-glutamate-co-γ-benzyl L-glutamate) Film. *J. Polym. Sci., Part B: Polym. Phys.* 1981, *19*, 1817-1826.

[40] Fujimori, A.; Hoshizawa, H.; Kobayashi, S.; Sato, N.; Kanai, K.; Ouchi, Y. Highly Ordered and Stable Layered "Polymer Nanosheets" Constructed with Amorphous Side Chains and Pi-Pi Stacking of Functional Groups in Ternary Comb Copolymers. *J. Phys. Chem. B* 2010, *114*, 2100-2110.

[41] Alfrey, T.; Price, C. C. Relative Reactivities in Vinyl Copolymerization. *J. Polym. Sci.* 1947, *2*, 101-106.

[42] Iwashita, H.; Hakozaki, S.; Kawaguchi, S.; Fujimori, A. Fine Structure of Multi-Particle Layered Organization for Organo-Modified Zirconium Dioxides Fabricated by the Langmuir-Blodgett Technique. *Trans. Mater. Res. Soc. Jpn.* 2012, *37*, 299-302.

[43] Fujimori, A.; Ninomiya, N.; Masuko, T. Structure and Mechanical Properties in Drawn Poly (l-lactide)/clay Hybrid Films. *Polym. Adv. Technol.* 2008, *19*, 1735-1744.

[44] Fujimori, A.; Arai, S.; Kusaka, J.; Kubota, M.; Kurosaka, K. Formation and Structure of Langmuir-Blodgett Films of Organo-Modified Aluminosilicate with High Surface Coverage. *J. Colloid Interf. Sci.* 2013, *392*, 256-265.

[45] Fujimori, A.; Kusaka, J.; Nomura, R. Formation and Structure of Organized Molecular Films for Organo-modified Montmorillonite and Mixed Monolayer Behavior with Poly (L-lactide). *Polym. Eng. Sci.* 2011, *51*, 1099-1107.

[46] Fujimori, A.; Araki, T.; Nakahara, H.; Ito, E.; Hara, M.; Ishii, H.; Ouchi, Y.; Seki, K. In-Plane X-Ray Diffraction and Polarized NEXAFS Spectroscopic Studies on the Organized Molecular Films of Fluorinated Amphiphiles with Vinyl Esters and Their Comb-Polymers. *Chem. Phys. Lett.* 2001, *349*, 6-12.

[47] Fujimori, A.; Sugita, Y.; Nakahara, H.; Ito, E.; Hara, M.; Matsuie, N.; Kanai, K.; Ouchi, Y.; Seki, K. Change of Molecular Packing and Orientation from Monolayer to Multilayers of Hydrogenated and Fluorinated Carboxylates Studied by In-Plane X-Ray Diffraction Together with NEXAFS Spectroscopy at C K-Edge. *Chem. Phys. Lett.* 2004, *387*, 345-351.

[48] Kumaki, J. Polystyrene Monomolecular Particles Obtained by Spreading Dilute Solutions on the Water Surface. *Macromolecules* 1986, *19*, 2258-2263.

[49] Hoshizawa, H.; Masuya, R.; Masuko, T.; Fujimori, A. Control of Orientation for Carbazole Group in Comb Copolymers Arranged by Method of Organized Molecular Films. *Polym. Adv. Technol.* 2007, *18*, 353-363.

[50] Hoshizawa, H.; Fujimori, A. Control of Arrangement for Carbazole Group in Fluorinated Comb Copolymers by Method of Organized Films. *Colloids Surf. A* 2008, *322*, 239-242.

[51] Fujimori, A.; Hayasaka, Y. Changes in Arrangement of Lamella and Fine Crystallite in Fluorinated "Crystalline" Transparent Fibers with Drawing. *Macromolecules* 2008, *41*, 7606-7615.

[52] Matsui, J.; Mitsuishi, M.; Aoki, A.; Miyashita, T. Molecular Optical Gating Devices Based on Polymer Nanosheets Assemblies. *J. Am. Chem. Soc.* 2004, *126*, 3708-3709.

[53] Mitsuishi, M.; Koishikawa, Y.; Tanaka, H.; Sato, E.; Mikayama, T.; Matsui, J.; Miyashita, T. Nanoscale Actuation of Thermoreversible Polymer Brushes Coupled with Localized Surface Plasmon Resonance of Gold Nanoparticles. *Langmuir* 2007, *23*, 7472-7474.

[54] Tanaka, H.; Mitsuishi, M.; Miyashita, T. Tailored-Control of Gold Nanoparticle Adsorption onto Polymer Nanosheets. *Langmuir* 2003, *19*, 3103-3105.

[55] Kunitake, T.; Okahata, Y.; Shimomura, M.; Yasunami, S.; Takarabe, K. Formation of Stable Bilayer Assemblies in Water from Single-Chain Amphiphiles. Relationship between the Amphiphile Structure and the Aggregate Morphology. *J. Am. Chem. Soc.,* 1981, *103*, 5401–5413.

[56] Fukuda, K.; Shibasaki, Y.; Nakahara, H.; Liu, M.-h. Spontaneous Formation of Polypeptides in the Interfacial Thin Films of Amphiphilic Amino Acid Esters: Acceleration of the Polycondensation and Control of the Structure of Resultant Polymers. *Adv. Colloid and Interface Sci.* 2000, *87*, 113–145.

[57] Kuhn, H.; Möbius, D.; Bucher, H. Spectroscopy of Monolayer Assemblies, in *Physical Methods of Chemistry*; Weissberger, A., Rossiter, B. W., Eds.; Wiley: New York, 1972; Vol. 1, Part IIIB, pp. 577-702.

[58] Appel, W.; Portale, G.; Wisse, E.; Dankers, P.; Meijer, E. W. Aggregation of Ureido-Pyrimidinone Supramolecular Thermoplastic Elastomers into Nanofibers: A Kinetic Analysis. *Macromolecules* 2011, *44*, 6776–6784.

[59] Wu, C.; Li, M.; Chi, S.; Jun, G. Laser Light Scattering Characterization of a Novel Polymer Nanofiber. *Macromolecules* 1998, *31*, 7553–7554.

[60] Huang, M. H.; Mao, S.; Feick, H.; Yan, H.; Wu, Y.; Kind, H.; Weber, E.; Russo, R.; Yang, P. Room-Temperature Ultraviolet Nanowire Nanolasers. *Science* 2001, *292*, 1897–1899.

[61] Bumm, L. A.; Arnold, J. J.; Cygan, M. T.; Dunbar, T. D.; Burgin, T. P.; Jones, L.; Allara, D. L.; Tour, J. M.; Weiss, P. S. Are Single Molecular Wires Conducting? *Science* 1996, *271*, 1705–1707.

[62] Park, K.; Sung, Y.; Han, S.; Yun, Y.; Hyeon, T. Origin of the Enhanced Catalytic Activity of Carbon Nanocoil-Supported PtRu Alloy Electrocatalysts. *J. Phys. Chem. B* 2004, *108*, 939–944.

[63] Liu, Y.; Li, H.; Tu, D.; Ji, Z.; Wang, C.; Tang, Q.; Liu, M.; Hu, W.; Liu, Y.; Zhu, D. Controlling the Growth of Single Crystalline Nanoribbons of Copper Tetracyanoquinodimethane for the Fabrication of Devices and Device Arrays. *J. Am. Chem. Soc.* 2006, *128*, 12917–12922.

[64] Johnston, E.; Bullock, S.; Uilk, J.; Gatenholm, P.; Wynne, K. Networks from α,ω-Dihydroxypoly(dimethylsiloxane) and (Tridecafluoro-1,1,2,2-tetrahydrooctyl) triethoxysilane: Surface Microstructures and Surface Characterization. *Macromolecules* 1999, *32*, 8173–8182.

[65] Klok, H.; Langenwalter, J.; Lecommandoux, S. Self-Assembly of Peptide-Based Diblock Oligomers. *Macromolecules* 2000, *33*, 7819–7826.

[66] Liu, F.; Prehm, M.; Zeng, X.; Tschierske, C.; Ungar, G. Skeletal Cubic, Lamellar, and Ribbon Phases of Bundled Thermotropic Bolapolyphiles. *J. Am. Chem. Soc.* 2014, *136*, 6846–6849.

[67] Komiyama, H.; Sakai, R.; Hadano, S.; Asaoka, S.; Kamata, K.; Iyoda, T.; Komura, M.; Yamada, T.; Yoshida, H. Enormously Wide Range Cylinder Phase of Liquid Crystalline PEO-*b*-PMA(Az) Block Copolymer. *Macromolecules* 2014, *47*, 1777–1782.

[68] Yabu, H.; Jia, R.; Matsuo, Y.; Ijiro, K.; Yamamoto, S.; Nishino, F.; Takaki, T.; Kuwahara, M.; Shimomura, M. Preparation of Highly Oriented Nano-Pit Arrays by Thermal Shrinking of Honeycomb-Patterned Polymer Films. *Adv. Mater.* 2008, *20*, 4200–4204.

[69] Fujimori, A.; Kaneko, Y.; Kikkawa, T.; Chiba, S.; Shibasaki, Y. Fabrication and Structure of "Polymer Nanosphere Multilayered Organization." *J. Colloid Interf. Sci.* 2014, *418*, 338–349.

[70] Joncheray, J. T.; Denoncourt, M. K.; Meier, M.; Schubert, S. U.; Duran, S. R. Two-Dimensional Self-Assembly of Linear Poly(ethylene oxide)-*b*-poly(ε-caprolactone) Copolymers at the Air-Water Interface. *Langmuir* 2007, *23*, 2423–2429.

[71] Kimizuka, N.; Kawasaki, T.; Hirata, K.; Kunitake, T. Supramolecular Membranes. Spontaneous Assembly of Aqueous Bilayer Membrane via Formation of Hydrogen Bonded Pairs of Melamine and Cyanuric Acid Derivatives. *J. Am. Chem. Soc.* 1998, *120*, 4094–4104.

[72] Blaaderen, A.; Ruel, R.; Wiltzius, P. Template-Directed Colloidal Crystallization. *Nature* 1997, *385*, 321–324.

[73] Rogach, A.; Susha, A.; Caruso, F.; Sukhorukov, G.; Kornowski, A.; Kershaw, S.; H.; Möhwald, Eychmüller, A.; Weller, H. Nano- and Microengineering: Three-Dimensional Colloidal Photonic Crystals Prepared from Submicrometer-Sized Polystyrene Latex Spheres Pre-Coated with Luminescent Polyelectrolyte/Nanocrystal Shells. *Adv. Mater.* 2000, *12*, 333–337.

[74] Leigh, A. D.; Murphy, A.; Smart, P. J.; Deleuze, S. M.; Zerbetto F. Controlling the Frequency of Macrocyclic Ring Rotation in Benzylic Amide [2]Catenanes. *J. Am. Chem. Soc.* 1998, *120*, 6458–6467.

[75] Kobayashi, A.; Dosen, M.; Chang, M.; Nakajima, K.; Noro, S.; Kato, M. Synthesis of Metal-Hydrazone Complexes and Vapochromic Behavior of Their Hydrogen-Bonded Proton-Transfer Assemblies. *J. Am. Chem. Soc.* 2010, *132*, 15286–15298.

[76] Akiyama, Y.; Kikuchi, A.; Yamato, M.; Okano, T. Ultrathin Poly(N-isopropylacrylamide) Grafted Layer on Polystyrene Surfaces for Cell Adhesion/Detachment Control. *Langmuir* 2004, *20*, 5506–5511.

[77] Muramatsu, M.; Akatsuka, K.; Ebina, Y.; Wang, K.; Sasaki, T.; Ishida, T.; Miyake, K.; Haga, M. Fabrication of Densely Packed Titania Nanosheet Films on Solid Surface by Use of Langmuir-Blodgett Deposition Method without Amphiphilic Additives. *Langmuir* 2005, *21*, 6590–6595.

[78] Zhao, J.; Jensen, L.; Sung, J.; Zou, S.; Schatz, C. G.; Van Duyne, P. R. Interaction of Plasmon and Molecular Resonances for Rhodamine 6G Adsorbed on Silver Nanoparticles. *J. Am. Chem. Soc.* 2007, *129*, 7647–7656.

[79] Yamada, K.; Nakano, T.; Okamoto, Y. Synthesis of Syndiotactic Poly(vinyl alcohol) from Fluorine-Containing Vinyl Esters. *Polym. J.* 1998, *30*, 641–645.

[80] Fujimori, A.; Araki, T.; Nakahara, H. Effect of the Atoms in Outermost Layers upon Friction Behaviors of Organized Molecular Films of Comb-Polymers Containing Fluorocarbons. *Chem. Lett.* 2000, *8*, 898–899.

[81] Klug, H. P.; Alexander, L. E. *X-Ray Diffraction Procedures*; John Wiley and Sons: New York, 1974.

[82] Vand, V.; Bell, I. P. A Direct Determination of the Crystal Structure of the β Form of Trilaurin. *Acta. Cryst.* 1951, *4*, 465–469.

[83] Yokoyama, D.; Park, Y.; Kim, B.; Kim, S.; Pu, Y.; Kido, J.; Park, J. Dual Efficiency Enhancement by Delayed Fluorescence and Dipole Orientation in High-Efficiency Fluorescent Organic Light-Emitting Diodes. *Appl. Phys. Lett.* 2011, *99*, 123303.

[84] Ghosh, D.; Chattopadhyav, N. Characterization of the Excimers of Poly(N-vinylcarbazole) Using TRANES *J. Lumin.* 2011, *131*, 2207–2211.

[85] Fujimori, A.; Kikkawa, T.; Meng, Q.; Shibasaki, Y.; Control of Fine Structure in "Polymer Nanosphere Multilayered Organization" and Enhancement of Its Optical Property. *Langmuir* 2015, *31*, 9177-9187.

In: Nanosheets and Nanospheres
Editor: Deborah Wright
ISBN: 978-1-53610-203-1

Chapter 4

OXYGEN ELECTROCHEMISTRY ON TWO DIMENSIONAL NANOSHEETS

Hengcong Tao and Zhenyu Sun*
State Key Laboratory of Organic-Inorganic Composites,
Beijing University of Chemical Technology, Beijing, China

ABSTRACT

Two-dimensional (2D) nanosheets have drawn great attention because of their interesting electronic and chemical properties since the isolation of graphene in 2004. This book chapter would summarize recent advances in the electrochemical oxygen-relative reactions, such as oxygen reduction reaction and oxygen evolution reaction by using electrocatalysts of graphene and its derivatives and also other 2D materials. From the mechanistic point of view, the link between structure and activity of these 2D nanosheets is discussed. Strategies that are effective to adjust electronic structure in these 2D materials and thereby catalytic activities are emphasized. Future opportunities and challenges in this field are outlined.

* Corresponding Author Email: sunzy@mail.buct.edu.cn.

1. INTRODUCTION

Due to rapidly growing energy demands and gradual consumption of fossil fuels that are limited in supply, the development of advanced technologies for efficient and clean energy conversion, especially in fuel cells, metal-air batteries and water splitting, is of substantial importance. The oxygen-relative reactions (oxygen reduction reaction, i.e., ORR and oxygen evolution reaction, i.e., OER) in electrochemical processes are the vital core of the above energy conversion and storage technologies. The ORR plays an important role in fuel cells and Zn/air batteries [1, 2]. The OER is essentially associated with clean energy systems of electrolytic/solar water splitting and rechargeable metal-air batteries by producing gaseous molecular hydrogen and oxygen [3, 4]. Note that both ORR and OER require efficient electrode materials to enhance sluggish kinetic processes and improve reaction efficiency. Electrocatalysis considers the impact of electrode materials and the state of electrode surfaces on the behavior of chemical-catalyzed steps being associated with electron transfer processes. It expedites transformations between electrical and chemical energy, displaying vital usefulness in energy storage and conversion devices. To date, the state-of-art electrode materials for oxygen-relative reactions are noble metals such as Pt and precious metal oxides (e.g., RuO_2 and IrO_2). They exhibit reasonable catalytic efficiency for the above electrochemical processes. Nonetheless, their scarcity and high cost remain problematic limiting large-scale applications. Therefore, searching for storage-abundant electrode materials with low cost and competitive performance is highly desirable.

Nanosheets refer to materials of two-dimensional (2D) nanostructure that are composed of less than 10 monolayers [5]. Dramatic changes in electronic properties can occur when the thickness of layered materials is reduced to a single or a few layers; for instance, extended bandgaps are attained due to quantum confinement and surface effect [6]. Equally importantly, large surface area is attainable when the bulk is thinned down to a few layers. Graphene is the mostly well-studied 2D material, comprising sp^2-hybridized carbon atoms arranged in a honeycomb lattice. An optical transmittance of 97.7%, impermeability to any gases, carrier mobility of up to 200000 $cm^2\ V^{-1}\ s^{-1}$ at room temperature and a Young's modulus of 0.5-1 TPa have been reported [5, 7, 8] These remarkable properties make graphene intriguing in electronics, composites, energy storage, to name but a few. The isolation of graphene directly from graphite in 2004 has led to an increasing interest in exploration of other 2D atomic crystals, including transition metal dichalcogenides (TMDs) [6, 9-11], metal oxides (e.g., MnO_2 and MoO_3) [12-15], hexagonal boron nitride (*h*-

BN) [16], graphitic carbon nitride [17], and MXenes (2D transition metal carbides and carbonitrides) [18].

Recently, the catalysis in oxygen electrochemistry reactions by using ultrathin 2D nanosheets has harvested snowballing research interest owing to their unique structure and electronic properties [19-21]. It is worth noting that a high number of exposed surface atoms in nanosheets can easily escape from the respective lattice, resulting in vacancy-type defects [22]. Such vacancy defects and accompanying structural disorder that usually forms in nanosheets, tend to lower the coordination state of surface atoms [19]. This leads to dangling bonds and high catalytic activities [10]. Importantly, manipulation of vacancy defects permits one to modify the electronic structure and thereby tune corresponding catalytic activities [10]. Additionally, 2D nanosheets with abundant exposed active sites are beneficial for fast interfacial charge transfer and facile electrochemical catalysis. Notably, the physical and chemical properties of 2D nanosheets can be tuned in diverse ways by using doping, and heterostructural and alloying strategies, making them an interesting platform for study in catalysis. Of diverse 2D materials, graphene and its derivatives, chalcagenides, nitrides, noble metals, and non-noble metal oxides bring new perspectives and prospects in oxygen-relative reactions.

2. Oxygen Reduction Reaction on Nanosheets

The ORR plays a key role in H_2/O_2 and other fuel cells as well as in Zn/air batteries. As is well known, by combining with a proton into water, the ORR pathway can be divided into a two-step/two-electron pathway with formation of H_2O_2 (in acidic medium) or HO_2^- (in alkaline medium) as the intermediate specie and a more efficient four-electron process to directly reduce O_2 into H_2O (in acidic medium) or OH^- (in alkaline medium) as follows [23, 24].

(1) Acidic aqueous solution:

$$O_2 + 4H^+ + 4e^- \rightarrow 2H_2O\,;\, E^0 = 1.229\text{ V vs. SHE}$$
$$O_2 + 2H^+ + 2e^- \rightarrow H_2O_2\,;\, E^0 = 0.695\text{ V vs. SHE}$$
$$H_2O_2 + 2H^+ + 2e^- \rightarrow 2H_2O\,;\, E^0 = 1.776\text{ V vs. SHE}$$

(2) Alkaline aqueous solution:

$$O_2 + H_2O + 4e^- \rightarrow 4OH^-; E^0 = 0.401 \text{ V vs. SHE}$$
$$O_2 + H_2O + 2e^- \rightarrow HO_2^- + OH^-; E^0 = -0.076 \text{ V vs. SHE}$$
$$HO_2^- + H_2O + 2e^- \rightarrow 3OH^-; E^0 = 0.878 \text{ V vs. SHE}$$

The significance of these 2- and 4-electron reduction pathways depends on targeted applications. The 4-electron direct pathway is highly preferred in fuel cell processes, while the 2-electron reduction pathway is used in industry for H_2O_2 production.

The ORR mechanisms are also influenced by electrode materials. Two mechanisms have been proposed for metal-free-catalyzed ORR. On a graphite electrode, the ORR proceeds by the following reaction mechanism [25].

$$O_2 \rightarrow O_{2(ads)}$$
$$O_{2(ads)} + e^- \rightarrow [O_{2(ads)}]^-$$
$$[O_{2(ads)}]^- \rightarrow O_{2(ads)}^-$$
$$O_{2(ads)}^- + H_2O \rightarrow HO_{2(ads)} + OH^-$$
$$HO_2(ads) + e^- \rightarrow HO_2^-{}_{(ads)}$$
$$HO_2^-{}_{(ads)} \rightarrow HO_2^-$$

On a pyrolytic graphite electrode, the first reaction was also proposed as follows.

$$O_2 + e^- \rightarrow O_{2(ads)}^-$$
$$2O_{2(ads)}^- + H_2O \rightarrow O_2 + HO_2^- + OH^-$$

Theoretical calculations by density functional theory (DFT) have been applied to investigate the ORR on metal catalysts [26]. Dissociative mechanism was suggested to be predominant in low current density regimes while associative mechanism plays a role in high current density regimes.

(1) Dissociative mechanism:

$$1/2O_2 + * \rightarrow O*$$
$$O* + H^+ + e^- \rightarrow OH*$$
$$OH* + H^+ + e^- \rightarrow H_2O + *$$

where * denotes a site on the catalyst surface. In this mechanism, no H_2O_2 is produced. O_2 adsorption breaks the O-O bond forming adsorbed atomic O, which further gains two electrons in two consecutive steps to generate water. Since there is no adsorbed O_2 on the catalyst surface, H_2O_2 is unlikely formed. This mechanism can be considered as a detailed form of the direct 4-electron pathway.

(2) Associative mechanism:

$$O_2 + * \rightarrow O_2 *$$
$$O_2 * + H^+ + e^- \rightarrow HO_2 *$$
$$HO_2 * + H^+ + e^- \rightarrow H_2O + O_2 *$$
$$O * + H^+ + e^- \rightarrow OH *$$
$$OH * + H^+ + e^- \rightarrow H_2O + *$$

This mechanism does not involve H_2O_2 either. Since adsorbed O_2 is absent, the O-O bond may not be broken in the following steps to yield H_2O_2.

Table 1. Overview of the onset potential and current density of oxygen related reactions on two-dimensional electrocatalysts

	Electrocatalyst	Onset potential /V	Current density /mA cm^{-2}	Electrolyte	Ref.
ORR	B-graphene	-0.05 V (vs. Ag/AgCl)		0.1M KOH	[27]
		0.83 V (vs. RHE)		0.3M KOH	[28]
	N-graphene	0.743 V (vs. RHE)		0.1M KOH	[29]
		~0.755 V (vs. RHE)		0.1M KOH	[30]
	N-porous carbon nanosheets	-0.076 V (vs. Ag/AgCl)	-5.79 mA cm^{-2} at -0.8 V (vs. Ag/AgCl)	0.1M KOH	[31]
	P-graphene	0.70 V (vs. RHE)	-4.18 mA cm^{-2} at 0.83 V (vs. RHE)	0.1M KOH	[32]
		0.80 V (vs. RHE)		0.1M KOH	[28]
		0.10 V (vs. Ag/AgCl)		0.1M KOH	[33]
	S-graphene	-0.09 V (vs. Ag/AgCl)	-6.99 mA cm^{-2} at -0.1 V (vs. Ag/AgCl)	0.1M KOH	[34]

Table 1. (Continued)

	Electrocatalyst	Onset potential /V	Current density /mA cm^{-2}	Electrolyte	Ref.
		-0.35 V (vs. Ag/AgCl)	-9.34 mA cm^{-2} at -0.30 V (vs. Ag/AgCl)	0.1M KOH	[35]
	Cl-graphene	-0.16 V (vs. Ag/AgCl)	-0.18 mA cm^{-2} at -0.80 V (vs. Ag/AgCl)	0.1M KOH	[36]
	Br-graphene	-0.15 V (vs. Ag/AgCl)	-0.28 mA cm^{-2} at -0.80 V (vs. Ag/AgCl)	0.1M KOH	[36]
	I-graphene	-0.14 V (vs. Ag/AgCl)	-0.40 mA cm^{-2} at -0.80 V (vs. Ag/AgCl)	0.1M KOH	[36]
	B/N-graphene	-0.03 V (vs. Ag/AgCl)		0.1M KOH	[37]
		0.82 V (vs. RHE)		1M $HClO_4$	[38]
	S/N-graphene	-0.15 V (vs. Ag/AgCl)	-10.0mA cm^{-2} at -0.25 V (vs. Ag/AgCl)	0.1M NaOH	[39]
		-0.12 V (vs. Ag/AgCl)		0.1M KOH	[40]
		-0.13 V (vs. Ag/AgCl)	-3.72mA cm^{-2} at -0.80 V (vs. Ag/AgCl)	0.1M KOH	[41]
	P/N-graphene	0.82 V (vs. RHE)		1M $HClO_4$	[38]
	B/N/ P-graphene	0.83 V (vs. RHE)		1M $HClO_4$	[38]
	N/P/ S-graphene	-0.192 V (vs. SCE)		0.1M KOH	[42]
	Graphene/ C_3N_4 nanosheets		-7.3 mA cm^{-2} at -0.40 V (vs. Ag/AgCl)	0.1M KOH	[43]
	Pt nanosheets	-0.08 V (vs. Ag/AgCl)		0.1M KOH	[44]
	Pt@Au/ graphene	-0.06 V (vs. Ag/AgCl)		0.1M KOH	[45]
	Ag/graphene	0.93 V (vs. RHE)		0.1M KOH	[46]
	MnO_2 nanosheets	-0.20 V (vs. Ag/AgCl)		0.1M KOH	[47]

	Electrocatalyst	Onset potential /V	Current density /mA cm^{-2}	Electrolyte	Ref.
	Au@MnO_2 nanosheets	0.92 V (vs. RHE)	-1.41 mA cm^{-2} at 0.9 V (vs. RHE)	0.1M KOH	[48]
	Ag-MnO_2/ graphene	0.94 V (vs. RHE)		0.1M KOH	[46]
	MoS_2/porous carbon nanosheets	-0.14 V (vs. Ag/AgCl) half-wave potential	-5.4 mA cm^{-2} at -0.60 V (vs. Ag/AgCl)	0.1M KOH	[49]
	$MoSe_2$	-0.02 V (vs. Ag/AgCl)	-10.22 mA cm^{-2} at -0.20 V (vs. Ag/AgCl)	0.1M KOH	[50]
	WSe_2/ graphene	-0.02 V (vs. Ag/AgCl)	-10.77 mA cm^{-2} at -0.20 V (vs. Ag/AgCl)	0.1M KOH	[50]
OER	γ-CoOOH nanosheets	0.93 V (vs. RHE)	50 mA cm^{-2} at 1.58 V	1M KOH	[51]
	Co_3O_4 nanosheets		342 mA cm^{-2} at 1.0 V (vs. Ag/AgCl)	1M KOH	[52]
	CoFe LDH	1.5 V (vs. RHE)	10 mAcm^{-2} at 325 mV (vs. Ag/AgCl) over potential	0.1M KOH	[53]
	ZnCo/ graphene		10 mA cm^{-2} at 300 mV (vs. Ag/AgCl) over potential	0.1M KOH	[54]
	NiFe/graphene		10 mA cm^{-2} at 230 mV (vs. Ag/AgCl) over potential	1M KOH	[55]
	Ni_3S_2 nanosheets		10 mA cm^{-2} at 260 mV (vs. Ag/AgCl) overpotential	1M KOH	[56]
	NiV LDH		10 mA cm^{-2} at 300 mV (vs. Ag/AgCl) overpotential	1M KOH	[57]
	Ni/N-graphene	0.32 V (vs. Ag/AgCl)		0.1M KOH	[58]

Table 1. (Continued)

	Electrocatalyst	Onset potential /V	Current density /mA cm^{-2}	Electrolyte	Ref.
OER and ORR	Co_9S_8/ graphene	0.941 V (vs.RHE) ORR 1.51 V (vs. RHE) OER	10 mA cm^{-2} at 0.409 V (vs. RHE) overpotential OER	0.1M KOH	[59]
	B-graphene	-0.20 V (vs. Ag/AgCl) ORR	-16.32 mA cm^{-2} at -0.60 V (vs. Ag/AgCl) ORR5 mA cm^{-2} at 0.9 V (vs. Ag/AgCl) OER	0.1M NaOH	[60]
	N/P-graphene	1.01 V (vs.RHE) ORR	-3 mA cm^{-2} at 0.86 V (vs. RHE) ORR -10 mA cm^{-2} at 1.57 V (vs. RHE) OER	0.1M KOH	[61]
	Co_3O_4/N-graphene	0.88 V (vs.RHE) ORR	-3 mA cm^{-2} at 0.82 V (vs. RHE) ORR -10 mA cm^{-2} at 1.54 V (vs. RHE) OER	0.1M KOH	[62]
	$NiCo_2S_4$@N/ S-graphene	0.85 V (vs.RHE) ORR	-3 mA cm^{-2} at 0.85 V (vs. RHE) ORR -10 mA cm^{-2} at 1.70 V (vs. RHE) OER	0.1M KOH	[63]
	$NiCo_2O_4$/ graphene	0.89 V (vs.RHE) ORR	-3 mA cm^{-2} at 0.72 V (vs. RHE) ORR -10 mA cm^{-2} at 1.69 V (vs. RHE) OER	0.1M KOH	[64]

2.1. Oxygen Reduction Reaction on Metal-Free 2D Catalysts

Metal-free materials with high efficiency, remarkable acidic durability and low-cost have gained heightened attention in ORR, initiating a practical goal of reducing our dependence on rare and expensive metallic catalysts. In this regard, graphene and its derivatives present many advantages. Graphene not only possesses large surface area (theoretically ~ 2630 $m^2 g^{-1}$ for graphene), affording a potentially high density of surface active sites [7], but also has outstanding electrical conductivity (facilitated electron transfer) and high mechanical strength and stability (tolerance to harsh operational conditions), which are beneficial when used as either the catalyst or catalyst support.

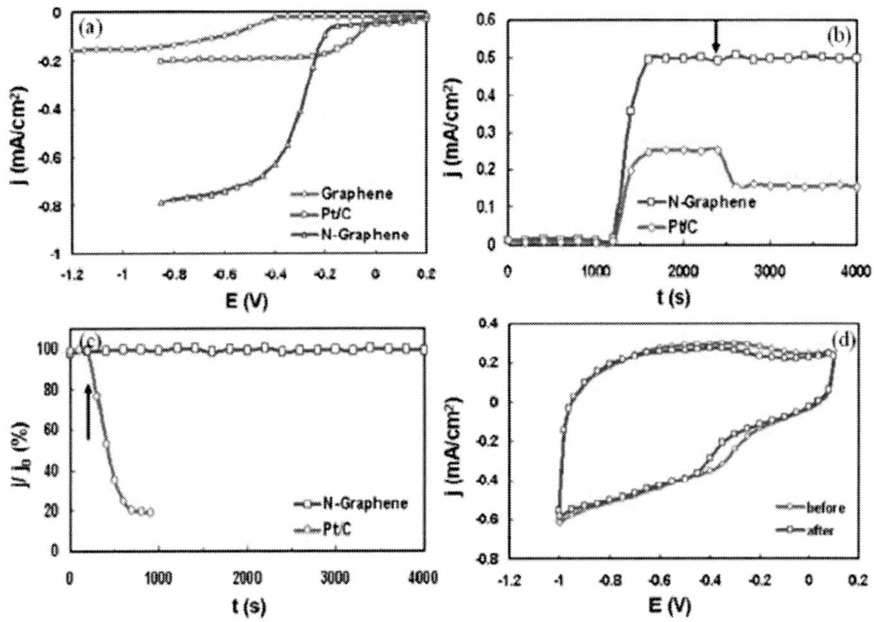

Figure 1. (a) RRDE voltammograms for the ORR in air-saturated 0.1 M KOH at the C-graphene electrode (red line), Pt/C electrode (green line), and N-graphene electrode (blue line) (b) Current density (j)/time (t) chronoamperometric responses obtained at the Pt/C (circle line) and N-graphene (square line) electrodes at -0.4 V in air saturated 0.1 M KOH. (c) Current (j)/time (t) chronoamperometric response of Pt/C (circle line) and N-graphene (square line) electrodes to CO. (d) Cyclic voltammograms of N-graphene electrode in air saturated 0.1 M KOH before (circle line) [60].

In general, graphene exhibits a very low density of states (DOS) at the Fermi level, and is catalytically inert at the basal plane probably resulting from its weak electronic coupling nature. Nevertheless, the local electronic structure can be effectively modified by chemical doping with heteroatoms that have a different electronegativity relative to C, such as nitrogen [65-68], boron [69-71], sulfur [35, 72-75], or phosphorus [29, 76], halogen [29, 76, 77], yielding a metal-free ORR catalyst. Relevant information regarding the type, level and configuration of heteroatom doping of graphene would be discussed in the following parts, while detailed aspects on the doping procedures and experimental conditions are beyond the scope of this book chapter, which can be found in earlier works [5, 65]. Note that the morphology and surface area of graphene are preserved regardless of such treatment, emphasis will thus be placed on the impacts of doping on the electronic properties of graphene and thereby on ORR performance. The electrocatalytic performance of doped graphene catalysts seems to rely on charge density, spin density and energy gap. DFT calculations indicated that different types of dopants and different doping sites in graphene could change the free energy diagram of oxygen reduction reaction pathways. The volcano plot has been established between experimentally measured ORR activity and theoretically predicted adsorption free energy of intermediates for graphene and its derivatives [78].

Doping of graphene with B requires a Gibbs free energy of 5.6 eV per atom (130 kcal mol^{-1}) and is more prone to occur than doping of graphene with N. In-plane doping results in sp^2 hybridization of B atoms in the carbon lattice [79-82]. The planar structure of graphene is hence retained to a great extent. Whereas charge polarization can take place between adjacent C (electronegativity of carbon: 2.55) and electron deficient B atoms (electronegativity of boron: 2.04). Meanwhile, the longer B-C bond length (~1.50 Å) than C-C bonds (1.42 Å) can lead to a small alteration of lattice parameters [83, 84]. By contrast, out-of-plane doping at carbon divacancy sites results in a tetrahedral-like BC_4 unit which induces a distortion of graphene's planar structure (with B in-plane and the carbons distorted out of plane) [83, 84]. Doping of a B atom in the carbon lattice allows one to open a band gap between the valence and conduction bands, thus tuning the electrical properties of graphene. A p-type behavior results from B doping with a downshift of the Fermi level towards the Dirac point. The bandgap increases with the increase of B concentration, and can be adjusted in the range of 0-0.54 eV *via* controllable doping [80, 82]. Of interest is that graphitic B-doping can enhance electrical conductivity by introducing more holes to the valence band of graphene, and increase the carrier concentration (~ 0.5 carriers per dopant) [82, 85]. The electron-deficient B atoms promote chemisorption of the negatively polarized O atoms and hence facilitate O-O cleavage in the ORR [86, 87]. Recently, boron-doped graphene was reported to be the best non-metal doped graphene electrocatalyst for ORR in alkaline media with an onset potential of 0.035 V [29]. A following DFT study performed by Valentin et al. showed an overpotential of 0.29 V that is necessary to initiate an overall downhill reduction process in acidic conditions (pH = 0) and the analogous overpotential of 0.28 V in alkaline solutions (pH = 14). The onset potential for ORR on B-doped graphene samples in alkaline electrolytes is calculated to be 0.05 V, in agreement with the experimental reported value [88]. Xia and co-workers synthesized boron doped graphene (B-G) by thermal-annealing graphite oxide (GO) in the presence of B_2O_3. These authors demonstrated that the ORR process catalyzed by the pristine graphene is a two-step two-electron pathway with the formation of intermediate HO_2^- ions. By contrast, the B-G, as in the case of the Pt electrode, exhibited a one-step process for ORR with an onset potential of about -0.05 V, which is about 100 mV more positive than that on pristine graphene [27].

N doping induces polarization in the carbon network because of the higher electronegativity of N (3.04) than C. A bandgap is opened upon N doping near the Dirac point by suppressing the nearby density of states, which imparts semiconducting properties in graphene [89-92]. Three primary N-bonding configurations have been suggested including basal plane quaternary N (central graphitic), pyrrolic N, and edge pyridinic N [87]. The sp^2 hybridized pyridinic and quaternary N do not influence the graphene structure whereas the sp^3 bonded pyrrolic N locally distorts the planar graphene structure [83]. The semiconducting property of N-doped graphene (N-G) is associated with the N bonding configuration. For quaternary N, three (sp^2) valence electrons of nitrogen form three σ-bonds with the neighboring carbon atoms, one ($2p_z$) electron is involved in π bonding, and the fifth electron is delocalized in the conduction band. The quaternary N causes an n-type doping effect by donating ~ 0.5 electron to the graphene lattice [89, 93], while pyridinic and pyrrolic nitrogen atoms that are located at defect sites lead to a p-type doping effect by sequestering electrons from the graphene conduction band network. A significant enhancement in carrier concentration was observed upon quaternary N doping, up to 2.6×10^{13} cm^{-2} (4 times higher than that of pristine graphene) even at a very small doping content of 0.6 atom%. N-G was first reported as an ORR catalyst by Dai and co-workers in 2010 [66]. The N-G cathode was shown to display larger current density, better tolerance to methanol crossover and stronger stability than the state-of-the-art Pt/C catalyst. Reasonably high ORR activity can be obtained, as demonstrated by Sun et al. even at a N doping level as low as 0.52 atom% [30]. Porous nitrogen-doped carbon nanosheets with high specific surface areas (nitrogen content 11.6 atom%, S_{BET} of 1077 $m^2\,g^{-1}$) were recently synthesized by Zhang et al. using g-C_3N_4 as a template and also as both a nitrogen source and porogen [31]. Such catalyst exhibited a high limiting current density of 5.79 mA cm^{-2}, and a low onset potential of -0.076 V (vs. Ag/AgCl) for ORR. Both quaternary N and pyridinic N moieties have been considered to accelerate ORR. This is consistent with previous results that quaternary N impacts the limiting current density, whereas the pyridinic N improves the onset ORR potential as reported by Ruoff et al. [67]. These authors proposed that the quaternary N can transfer electron density to the delocalized π system (n-type doping), resulting in enhanced nucleophile strength for neighbouring carbon rings, whereby promoting the O_2 adsorption, and thus favoring the ORR in alkaline electrolytes [67, 87]. Several groups postulated that pyridinic N appears as an active site for enhancing the ORR activity of N-G. The ORR active sites were further explored by using newly designed graphite model catalysts with well-defined π conjugation and well-controlled N-doping [65]. It was concluded that those carbon atoms being adjacent to the pyridinic N with Lewis basicity are most probably the active sites of ORR under acidic conditions, which facilitate the adsorption of oxygen molecules as the initial step.

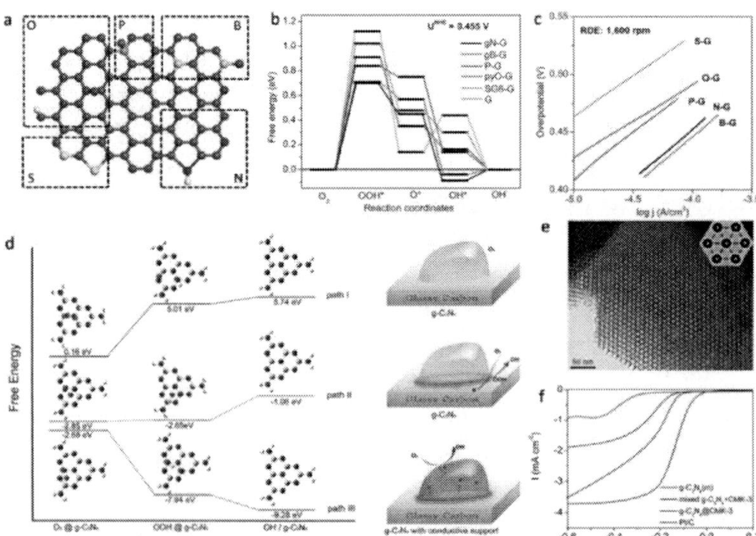

Figure 2. (a) Typical atomic configuration of different types of dopants at different doping sites in the graphene matrix. (b) Free energy diagram of different heteroatom-doped graphene at the equilibrium potential U_0. (c) Experimentally determined Tafel plots for different catalysts. (d) Free energy diagram and optimized configurations of adsorbed species on the g-C_3N_4 surface with zero, two and four electrons participation demonstrated as paths I–III. (e) Typical TEM image and (f) ORR performance of the synthesized g-C_3N_4@CMK-3 composites [31].

Substitutional P doping can result in sp^3 hybridization of carbon atoms, yielding a tetrahedral-like distorted structure with three proximal carbon atoms. The P-C bond length (1.77 Å) is longer than the C-C bonds, rendering the protrusion of P atoms from the graphene plane by around 0.09 Å. Provided that the electronegativity of P (2.19) is lower than that of C (2.55), the polarity of the C-P bond is opposite to that of the C-N bond [76, 83]. In contrast to the 0.5 electrons that quaternary N transfers to the π system, P only donates around 0.2 electrons (Hirshfeld population analysis) [83, 94]. P-doped bilayers exhibit a stronger n-type behavior and higher mobility albeit with weaker electron donation than N-doped graphene [83]. Additionally, the n-type behavior resulting from P doping is more stable in an oxygen environment than the n-type behavior caused by N doping. P atoms doped into the graphene structure promote the ORR to give a more positive onset potential and larger current density than undoped graphene [32, 33]. Nonetheless, the ORR mechanism catalyzed by P-doped graphene remains a subject of ongoing discussion [76]. Liu et al. postulated that O_2 was firstly transformed to peroxide, which was partially reduced to water [33]. By contrast, Wu and co-workers proposed a four-electron pathway of ORR to form H_2O directly [95]. Based on DFT calculations, Yang et al. inferred that the ORR on P-G proceeds firstly via a two-electron step yielding an OOH intermediate, followed by a four-electron pathway to cleave the O-O bond of OOH. The second OH reduction to H_2O requiring a large barrier of 0.88 eV is possibly the rate-limiting step. P atoms and their neighboring C atoms are postulated to be preferable adsorption sites and active centers of ORR. The P atoms can also function as a bridge promoting electron transfer from C to adsorbates [96]. Hou et al. demonstrated a thermal annealing method for the synthesis of P-G using graphite oxide and triphenylphosphine as carbon and phosphorus sources, respectively. A significantly high ORR onset potential of 0.92 V (vs. RHE) was observed for the resultant P-G, which is much more positive than that of graphene (0.82 V. vs RHE) and close to that of the commercial Pt/C (0.95 V vs. RHE) [32].

Substitutional S (VI group element) doping of graphene results in two different forms of sulfide, i.e., C-S-C (thiophene) and oxidized sulfur, i.e., C-SO_x-C (x = 2, 3 or 4) [35, 72]. The length of the C-S bond (1.78 Å) is about 25% longer than that of the C-C bond. This implies that each S atom is approximately 1.1 Å out of the graphene plane [97]. Polarization would not take place in the C-S bond due to similar electronegativities of S (2.58) and C (2.55). Nevertheless, the highest occupied molecular orbital (HOMO) and lowest unoccupied molecular orbital (LUMO) of graphene can be strongly polarized by edge-sited S dopants, thus significantly increasing the spin density to afford high catalytic activity for ORR. The ORR active sites were suggested to be those C atoms located at zigzag edges or being adjacent to SO_x dopants [72]. The ORR mechanism on S-G depends on the doping structure. A two-electron transfer occurs *via* the adsorption of OOH species on the S atom followed by breaking of the bonds between sulfur and oxygen atoms to yield two H_2O_2 molecules. With respect to those active sites of carbon atoms with large spin or charge density, a four-electron transfer likely takes place [72, 73]. Yang and co-workers reported the fabrication of sulfur-doped graphene (S-G) by directly annealing graphene oxide and benzyl disulfide in argon. The peak potential of the ORR for the S-G shifted positively to -0.32 V and its oxygen-reduction current was (around 123 μA) remarkably higher than that of pristine graphene (82 μA) [35]. Alternatively, Kim et al. demonstrated the formation of S-G by thermal treatment of exfoliated graphene under CS_2 gas flow. The resulting S-G exhibited an onset potential around -0.11 V, in contrast to that of -0.15 V for pristine exfoliated graphene [34].

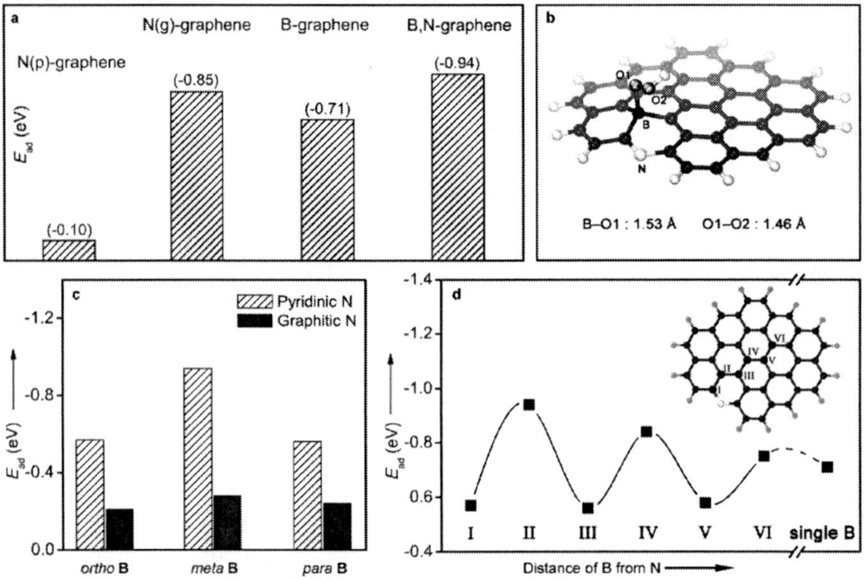

Figure 3. a) Energies of HO_2 adsorption (E_{ad}) on N- and/or B-doped graphene. N(p) and N(g) indicate pyridinic and graphitic N bonding, respectively. b) Optimized configuration of HO_2 adsorbed on B,N-graphene. c) E_{ad} on various B,N-graphene models with pyridinic or graphitic N groups. d) E_{ad} on various B,N-graphene models with B active sites as a function of the distance to a pyridinic N atom (sites I–VI as marked in inset). B-graphene with a single B atom was considered as a reference with an infinitely large distance [69].

Halogen atoms exhibit both electron withdrawing and donating effects due to their higher electronegativity ($\delta_F = 4.0$, $\delta_{Cl} = 3.0$, $\delta_{Cl} = 2.8$, $\delta_I = 2.5$) than C atoms, and the existence of lone pair of electrons, respectively. The behavior of modification in the electronic structure of graphene strongly depends on the halogen atoms. Fluorine with the largest electronegativity, tends to induce charge delocalization, altering the charge distributions in the G framework, and thus facilitating oxygen adsorption, which would further enhance its electrocatalytic activity towards ORR. Fluorinated graphene has a wide band gap of 3.1 eV, while chlorinated graphene has a much smaller band gap of 0.9 eV and brominated graphene has almost zero band gap. In addition, the band gap of F-doped graphene $(C_1H_xF_{1-x-\delta})_n$ can be readily tuned by manipulating the level of fluorination. On the basis of the occupation of the lowest-unoccupied molecular orbital level of O_2, the efficiency of charge transfer between the halogen and adsorbed O_2 follows the atomic size in the order of Cl $<$ Br $<$ I [36]. The diffusion-limiting currents at -0.8 V for pristine graphite, Cl-G, Br-G, I-G and Pt/C electrodes are -0.09, -0.18, -0.28, -0.40 and -0.30 mA, respectively. These current values for the Cl-G, Br-G and I-G are about 2, 3.1 and 4.4 times higher than that of the pristine graphite and 60%, 93% and 133% that of the Pt/C. Iodine-doped graphene obtained by annealing graphene oxide and iodine at 500-1100°C in argon showed kinetic-limiting current density of 9.21 mA cm^{-2} at -0.30 V vs Ag/AgCl. Good durability for methanol was also observed, which decreased very slowly showing a high relative current of 83.1% after 18000 s. In contrast, the Pt/C electrode exhibited a gradual decrease, with a current loss of approximately 70.5% after 18000 s.

Substitution of some carbon atoms with multiple heteroatoms like B/N [37, 98-100], S/N [39-41, 100, 101], B/P [28], P/N [100], B/N/P [38, 100, 102], N/P/S [42, 103], N/Fe, and N/S/Fe can effectively modify the electronic structure of graphene (or carbon nanosheets), creating novel properties or inducing "collaborative" effects [104, 105]. Likely, the co-doping with two elements that have reverse electronegativity to that of C, e.g., B and N, could bring about unique electron-donor properties due to synergistic coupling effects, and drastically accelerate the ORR. For instance, a four-electron pathway with positive ORR onset potential was shown by Qiao et al. upon co-doping of graphene with N and B dual atoms provided the absence of BN bonding [98]. The resulting B/N-G showed high kinetic current density (13.9 mA cm^{-2}) exceeding the sum of kinetic current densitiesderived for B-G (3.48 mA cm^{-2}) and N-G (5.54 mA cm^{-2}). Two possible aspects account for such enhancement. On the one hand, B/N-G exhibits higher OOH_{ads} adsorption energy than either quaternary N-G, pyridinic N-G, or B-G. On the other hand, B and N co-doping can promote the donation of electrons to neighboring B atoms, which results from the N-induced polarization of C $2p_z$ orbital located between N and B dopants. A B-N bond may be formed when the edge-plane C is replaced by N. Such configuration is, however, undesirable which can bring about inferior graphite-like ORR activities probably due to the complementary effect between n- and p-type of dopants [106]. Similar collaborative effects of S and N co-doping on ORR were also observed. Furthermore, co-doping can tune the electronic properties of graphene inducing additional electron-acceptor and electron-donor states, thus generating more active sites for ORR. P/N co-doped graphene was demonstrated to have an ORR onset potential of around 0.87 V. The mass activity of this dually doped graphene is 0.80 mA mg^{-1} being 1.8 times higher than that of N-G [107]. This activity enhancement may stem from the increased asymmetry of spin density or electron transfer in the basal plane of graphene, and the decrease in the energy gap between the HOMO and the LUMO of graphene upon P and N co-doping. Doping of graphene with a second phase, besides N, may impact the bonding configuration. In this regard, Xu et al. showed that in contrast to the formation of three C-N bonding conformations by individual N-doping, S/N co-doping of graphene could facilitate the evolution of (graphitic) quaternary N [39]. Based on first-principles calculations, Kim et al. showed that co-doping of N-G with a B or P atom allows electron transfer from the pyridinic N to the graphene π^* states, inducing the conversion of a p-type behaviour into an n-type behaviour. While only a bipolar type conversion can be attaiined for nitrilic or pyrrolic N, taking into account that the nitrilic and pyrrolic defect states are energetically located further away from E_F and cannot strongly hybridize with the graphene π states. Fe, N co-doped graphene was prepared by heating the mixture of graphene oxide and Fe-pentaethylenehexamine complex at 900°C in Ar for 45 s. The modified graphene acted as an efficient electrocatalyst for the ORR with an onset-potential of 0.85 V (vs. RHE) in acidic solutions [108].

Although interesting catalytic activities of heteroatom-doped graphene for ORR have been demonstrated, it remains a challenge to uniformly dope graphene by heteroatoms especially at high concentrations in a controllable manner, and to develop highly efficient catalysts. In addition, there has been no consensus on the origin of the ORR activity of graphene-based materials. From these scenarios, future works need to develop fine control in doping level and configuration, and identify ORR intermediates by using in-situ characterization techniques (such as spectroscopy). As an isoelectronic analogue of graphene, hexagonal boron nitride (*h*-BN) monolayer is another widely studied 2D material, which exhibits excellent mechanical properties and high thermal conductivity. In contrast to the semimetallic graphene with a zero energy gap, pristine *h*-BN nanosheet is a semiconductor with a wide band gap (5.0-5.9 eV) and shows notable oxygen reduction reactivity [109]. By means of spin-polarized DFT computations, Chen et al. showed that substitutional doping of C at the N site of a *h*-BN monolayer facilitated an OOH hydrogenation pathway in the ORR with free energy barrier of 0.61 eV, being lower than that of Pt-based catalyst (0.79 eV) [110, 109, 108]. These computation results suggested that carbon-doped *h*-BN nanosheets show potential as an alternative non-Pt ORR catalyst. Graphene/hexagonal boron nitride (G/*h*-BN) heterostructures have been investigated in the ORR since they were successfully synthesized by Lou and co-workers [105]. The C–N interfaces of G/*h*-BN nanoribbons were shown to be a source of active sites for the ORR based on spin-polarized DFT-D calculations. The graphene/*h*-BN (7,7) nanoribbon features the best activity and efficiency among different sized *h*-BN nanoribbons [111].

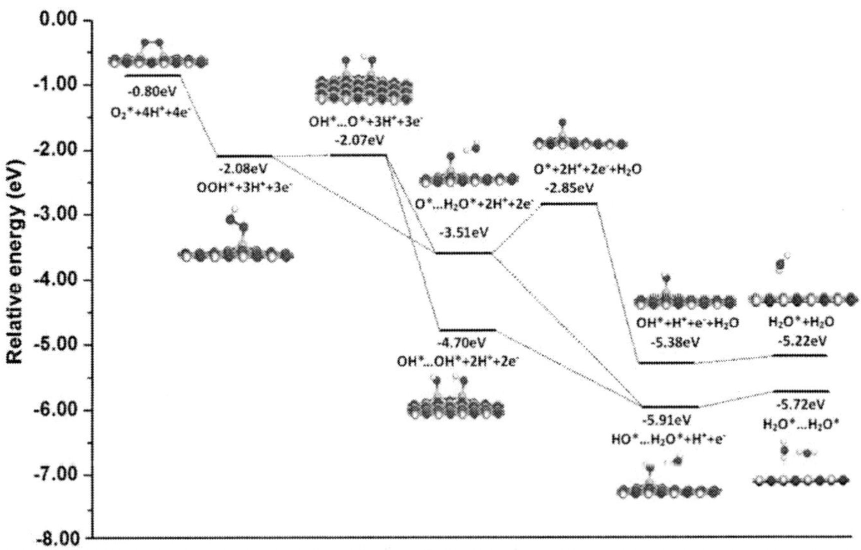

Figure 4. Reaction mechanism of water formation on the heterointerface of the graphene/*h*-BN (7,7) nanoribbon [111].

Because of its high nitrogen content, graphitic-carbon nitride (*g*-C_3N_4) may provide more active reaction sites than other N-carbon materials when used as a metal-free ORR electrocatalyst [112]. However, those active sites of *g*-C_3N_4 are limited to very narrow zones of the electrode-electrolyte-gas three-phase boundaries because of its poor electron transfer efficiency, which leads to unsatisfactory ORR performances. An effective way to increase the number of electrons accumulated on the *g*-C_3N_4 surface, and thereby extend the concentration of active sites is by introducing a highly conductive support such as carbon nanomaterials. For instance, based on theoretical predictions, *g*-C_3N_4@ carbon composite was reported to exhibit competitive catalytic activities (11.3 mA cm^{-2} of kinetic-limiting current density at -0.6 V vs. Ag/AgCl) and surprisingly high catalytic efficiency (nearly 100% of four-electron ORR process selectivity) [113]. Müllen et al. demonstrated an effective and controllable CVD approach for the fabrication of graphene-based C_3N_4 nanosheets [43]. The resulting graphene/C_3N_4 nanosheets exhibited a four-electron transfer pathway with a kinetic current density of up to 7.3 mA cm^{-2}. This is in stark contrast with those C_3N_4 nanosheets without graphene which showed a typical two-electron transfer pathway for the ORR with low current densities (J_k = 0.85-2.16 mA cm^{-2} at -0.40 V vs. Ag/AgCl).

2.2. Metal-Based 2D Catalysts for ORR

2D nanosheets of noble metals (Pt, Au, Ag, Pd, etc.), non-noble metal oxides (Co_3O_4, MnO_2, Fe_3O_4, Fe_2O_3 NiO, etc.) and metal chalcogenides (MoS_2, MoSe, ZnS, WSe_2, Co_8Se_9, etc.) are also the subject of research interest for the ORR. 2D hybrids combining two or more individual compositions have also been explored to enhance catalytic properties and decrease catalyst cost.

Pt has been the mostly well-studied ORR electrode since the 1960's, and is still the most active one. In particular, Pt nanosheets were shown to be ca. 125-fold significantly more active than commercial Pt/C [44]. However, the prohibitively high cost of Pt together with its limited abundance hampers practical applications in fuel cells. To solve this issue, an effective way is to support Pt on a conductive substrate with high surface area such as graphene. As an example, Jiang et al. showed that Pt monatomic wire supported on the zigzag edges of graphene nanoribbons can act as an efficient ORR electrocatalyst. The rate-determining step was found to be the reduction of O_2* to OOH* with maximal barrier of 0.41 eV at 1.23 V. Although pure Pt possess best activity for ORR, pure Pt is unsteady in fuel cells under acidic media, resulting in deactivated facets and degradation of catalytic activities. To improve the stability of Pt in harsh operating conditions, noble metals (Au, Ag, Pd, etc.) and non-precious metals (Ni, Fe, Co, Cu, Mn, etc.) have been incorporated into the Pt skeleton by rationally designed solution-based methods.

A novel hybrid consisting of bimetallic Pt@Au nanorods that were uniformly dispersed on nitrogen-doped graphene (Pt@Au/N-G) was reported by Wang and co-works. The resultant Pt@Au/N-G has significantly high ORR activity, which is 5-fold greater than that of commercial 20% Pt/C catalyst at -0.8 V vs Ag/AgCl, and enhanced electrocatalytic stability as well as superior methanol tolerance than the Pt/C [45].

MnO_2 nanosheets were prepared by mixing $KMnO_4$ solution and graphene oxide suspension followed by thermal treatment at 60℃ for 15h. These MnO_2 nanosheets exhibited an oxygen reduction peak at -0.29 V (vs. Ag/AgCl), in contrast to the reduction peak at -0.40, -0.40 and -0.44 V (vs. Ag/AgCl) for α-MnO_2 NPs, β-MnO_2 NPs and commercial MnO_2, respectively [47]. Liu et al. demonstrated that AuNP@MnO_2 nanosheets have large current density of up to -1.41 mA cm^{-2} at 0.9 V (vs. RHE), which is two-fold higher than that of commercial Pt/C catalyst (-0.82 mA cm^{-2}). The potential at current density = 2 mA cm^{-2} for AuNP@MnO_2 nanosheets is 0.88 V that is 30 mV higher than that of the Pt/C [48].

Silver is regarded as a promising candidate to replace Pt because of its relatively high electrochemical activity and long durability toward ORR in alkaline media. Ag/graphene-based materials, such as Ag/graphene and Ag-MnO_2/graphene, have been synthesized and extensively studied as eletrocatalysts for the ORR in alkaline media. By using poly diallyldimethylammonium chloride (PDDA) as a stabilizer, Sun and co-workers managed to synthesize reduced graphene oxide (rGO) supported Ag nanoparticles (NPs) of 10 nm in diameter. The average number of electrons transferred on this composite is 3.94 in the potential range from 0.2 to 0.5 V. Though its half-wave potential is slightly lower than that of Pt/C (0.80 V vs. RHE), its onset potential is comparable to that of Pt/C (0.93 V vs. RHE) [46]. In an analogous manner, Pd–Pt bimetallic NPs were anchored on rGO with the aid of PDDA [114]. The smaller lattice parameter of the Pd-rich core induced compressive strain on the outer shell such that the d-band center of the Pt-rich shell downshifted, weakening the adsorption energy of the oxygenated species, and thus enhancing the ORR kinetics of the Pd-Pt/rGO catalyst.

MoS_2 is inherently not an exciting ORR catalyst, while doping of MoS_2 nanosheets by heteroatoms such as P with a lower electronegativity than that of S, is beneficial to the preferential absorption of oxygen molecules on the surface, promoting the ORR from a two-electron process to a four-electron process. Both onset and half-wave potentials are 160 mV more positive than the pristine form, with an increase of the current density up to 7-fold. MoS_2/porous carbon nanosheet heterostructures, as recently demonstrated by Feng and co-workers, were able to efficiently catalyze the ORR with a more positive half-wave potential (-0.14 V vs. Ag/AgCl) compared to MoS_2-free porous carbons (-0.248 V vs. Ag/AgCl) [49]. Analogously, $MoSe_2$/graphene and WSe_2/graphene nanosheets were reported as efficient electrocatalysts for the ORR with an onset potential of -0.02 V (vs. Ag/AgCl) (close to that of Pt/C, 0.02 V vs. Ag/AgCl) and a high kinetic density of 10.22 mA cm^{-2} for $MoSe_2$/graphene and 10.77 mA cm^{-2} for WSe_2/graphene at -0.20 V (vs. Ag/AgCl). Interestingly, Song et al. demonstrated ultrathin MoS_2 sheets incorporated with high-electronegativity (3.44) O atoms by annealing of ammonium molybdate and thiourea using g-C_3N_4 as a sacrificial template followed by treatment with H_2O_2. This O incorporation may polarize those molybdynum atoms situated at the edges of MoS_2 nanosheets, inducing positive charges to absorb oxygen molecules from the electrolyte [50]. Meanwhile, the incorporated oxygen atoms alter the electronic structure of the nanosheets, causing a decrease in band gap and an increase in electrical conductivity. A substantial enhancement in ORR activity was attained with a 7.8-fold current increase, and 180 and 160 mV positive shifts in onset and half-wave potentials, respectively, compared to pristine MoS_2.

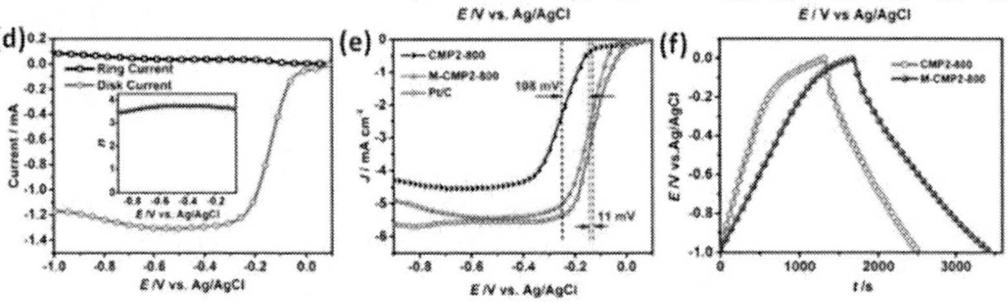

Figure 5. (a) The ORR process taking place at the surface of the M-CMPs-T hybrids in alkaline condition. (b) CV curves in N_2- and O_2-saturated 0.1m KOH. (c) LSV curves for M-CMP2-800 at different rotation rates in O_2-saturated 0.1 M KOH at 5 mV s^{-1}. The inset shows the Koutecky-Levich (K-L) plots. (d) RRDE curve for M-CMP2-800 for a rotation speed of 1600 rpm (Inset: calculated electron transfer number (n) against the potential). e) LSV curves for CMP2-800, M-CMP2-800, and Pt/C at 1600 rpm and a scan rate of 5 mV s^{-1}. (f) Galvanostatic charge/discharge curves of CMP2-800 and M-CMP2-800 at a current density of 0.2 A g^{-1} [95].

3. OXYGEN EVOLUTION REACTION ON 2D CATALYSTS

The oxygen evolution reaction (OER) is of significance because it is coupled with key clean energy systems such as electrolytic/solar water splitting and rechargeable metal–air batteries. As we know, the OER pathway is just a reverse version to that of the ORR pathway: in the ORR process, O_2 is reduced into H_2O or OH^-, while in the OER, H_2O or OH^- is oxidized to O_2. The mechanism of OER is very sensitive to the structure of electrode surface; different materials or one material with different facets can exhibit various reaction mechanisms. The generally accepted overall reaction pathways for OER involve four discrete electron transfer steps as follows:

(1) Acidic aqueous solution:

$$* + H_2O \rightarrow * OH + H^+ + e^-$$
$$* OH \rightarrow * O + H^+ + e^-$$
$$* O + H_2O \rightarrow OOH * + H^+ + e^-$$
$$* OOH \rightarrow * O_2 + H^+ + e^-$$
$$* O_2 \rightarrow * + O_2$$

Overall reaction: $2H_2O \rightarrow 4H^+ + 4e^- + O_2$

(2) Alkaline aqueous solution:

$$* + OH^- \rightarrow * OH + e^-$$
$$* OH + OH^- \rightarrow H_2O + * O + e^-$$
$$* O + OH^- \rightarrow * OOH + + e^-$$
$$* OOH + OH^- \rightarrow * O_2 + H_2O + e^-$$
$$* O_2 \rightarrow * + O_2$$

Overall reaction: $4OH^- \rightarrow O_2 + 4e^- + 2H_2O$

The OER is limited either by the strong adsorption of OOH_{ads} or the weak adsorption of O_{ads}. The difference in activation state binding free energy between activated O* and OH* has been used as a catalytic gauge for the OER on metal oxide catalysts to construct volcano-type plots. It predicts that an optimal, moderate binding energy of the OER intermediates leads to a high catalytic activity. The catalyst performance is usually assessed by the overpotential required to attain a current density of 10 mA cm^{-2} per geometrical area, which is close to the current density expected for a 10% efficient solar-to-fuels conversion device.

The first study of OER electrocatalysts dated back to the 1960s focusing on single noble metals. The experimentally established overpotential sequence for the OER in acidic solutions is: Ru<Ir<Pd<Rh<Pt. Recently, 2D materials and 2D nanosheet-supported composites have shown potential catalysis for the OER due to their high electrochemical surface areas and large number of active sites. Thanks to their earth abundant nature, environmental benignity, and unique 3d electron number and special e_g orbitals, nickel-based compounds have been discovered as good OER catalyst candidates. Great efforts have been devoted to studying the OER properties of Ni-based structures, spanning from nickel oxides, nickel-hydroxides, Ni based layer double hydroxides to nickel sulfides. For instance, Ni-O, Ni-N, Ni-Fe layered double hydroxide (LDH), Ni-S and Ni-Co LDH nanosheets were demonstrated to be robust and highly active catalysts for water oxidation. Hu et al. showed that the overpotentials were decreased by 40-54 mV due to the increase in active site density and conductivity upon exfoliation of Ni-Fe and Ni-Co LDHs. The catalytic activity of the Ni-based LDH nanosheets has the order of Ni-Fe>Ni-Co, both of which are more active than Co-Co LDH nanosheets. Metallic nitride nanosheets, such as Ni_3N, were also suggested to be a promising OER electrocatalyst with superior performance than their bulk counterparts. The activity enhancement likely originates from their high carrier concentration and disordered structure [115]. High-index faceted Ni_3S_2 nanosheet arrays were recently reported as highly active and stable electrocatalysts for water splitting. These arrays supported on nickel foam afforded a small overpotential of 260 mV (at a current density of 10 mA cm^{-2}), ~100% Faradaic yield and remarkable stability for >200 h for the OER under basic media [56].

Xie and co-workers showed that atomically-thin porous Co_3O_4 nanosheets exhibited an electrocatalytic current of up to 342 mA cm^{-2} at 1.0 V (vs. Ag/AgCl), about 50- and 30-times higher than those, respectively, of the bulk material and commercial 20 wt% Pt/C [21, 52]. A great fraction of coordination-unsaturated surface Co^{3+} atoms served as active sites to promote the OER process. These authors further found that $NiCo_2O_4$, $NiLa_2O_4$, $ZnCo_2O_4$ and Co_3O_4 nanosheets that are rich in oxygen vacancies exhibited lower overpotential, smaller Tafel slopes, higher current densities and larger TOF values than those samples that are poor in oxygen vacancies, pointing to the important role of oxygen vacancies in those OER processes [54, 116].

Theoretical investigations indicated that oxygen vacancies confined in ultrathin nanosheets could reduce the adsorption energy of H_2O, leading to enhanced OER efficiency. γ-CoOOH nanosheets of 1.4 nm in thickness were recently shown to be an efficient electrocatalyst for water oxidation, exhibiting a large mass activity of 66.6 A g^{-1} which is 20 and 2.4 times higher than that of bulk γ-CoOOH and the benchmarking IrO_2 electrocatalyst, respectively [51]. These authors inferred that the half-metallic nature of ultrathin nanosheets greatly improved the electrophilicity of H_2O and facilitated the interfacial electron transfer between Co ions and adsorbed -OOH species to yield O_2, attaining a high electrocatalytic activity for the OER.

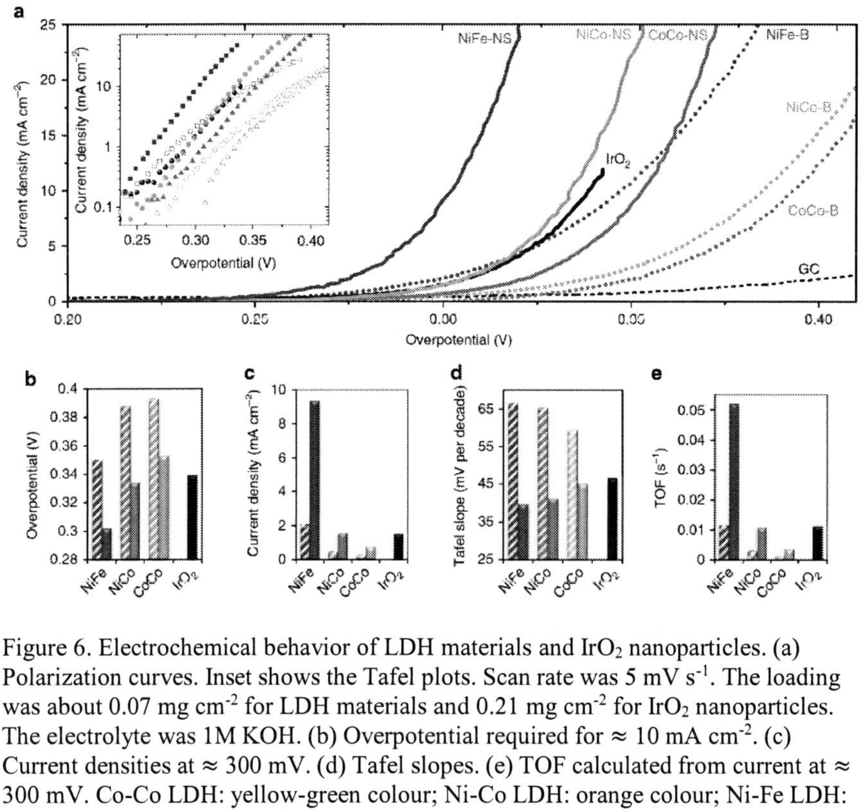

Figure 6. Electrochemical behavior of LDH materials and IrO_2 nanoparticles. (a) Polarization curves. Inset shows the Tafel plots. Scan rate was 5 mV s^{-1}. The loading was about 0.07 mg cm^{-2} for LDH materials and 0.21 mg cm^{-2} for IrO_2 nanoparticles. The electrolyte was 1M KOH. (b) Overpotential required for ≈ 10 mA cm^{-2}. (c) Current densities at ≈ 300 mV. (d) Tafel slopes. (e) TOF calculated from current at ≈ 300 mV. Co-Co LDH: yellow-green colour; Ni-Co LDH: orange colour; Ni-Fe LDH: purple colour. Dash lines, hollow symbols and patterned columns: bulk LDHs (B); Solid lines, symbols and columns: exfoliated LDHs nanosheets (NS); Black lines, symbols and columns: IrO_2 nanoparticles; GC: glassy carbon electrode [116].

N-doped graphene combined with transition metals (Ni), transition metal oxides, such as Co_3O_4, non-metal species including N-doped carbon nanotubes, and graphitic C were also demonstrated [117] as robust and active electrocatalysts for the OER [55]. For instance, Qiao et al. utilized N-doped graphene film as a scaffold to confine Ni nanoparticles, resulting in surprisingly high catalytic activity for OER. This hybrid catalyst exhibited an overpotential of 0.32 V, approaching that of the state-of-the-art IrO_2 catalyst (0.27 V) [58]. A dual-active-site reaction mechanism was proposed. On the one hand, Ni nanoparticles in Ni/N-G were partly oxidized into NiOOH giving rise to NiOOH/Ni complex active centers, which promote the oxidation of OH^- into molecular oxygen. These Ni nanoparticles can also improve charge transport of the electrodes. On the other hand, those positively charged carbon atoms due to N doping can adsorb OH^- through electrostatic forces to facilitate water dissociation. Bonding between Ni and N-doped graphene, i.e., Ni-O-C or Ni-N-C, may provide additional active centers in that Ni-O-C and Ni-N-C can promote the oxidation of OH^- into peroxide followed by peroxide decomposition into molecular oxygen.

Of particular interests are metal-free electrocatalysts for OER with low cost and earth abundance. Following the first work using an organic compound as a metal-free OER catalyst [118], N-doped graphite was demonstrated to efficiently catalyze the OER with an overpotential of 0.38 V at 10 mA cm^{-2} [119]. The pyridinic N and/or graphitic N species are postulated to be active sites for the OER. P doping of graphene was reported to lower the charge energy needed by 0.25 eV [120], while B-doping of graphene decreased the oxygen evolution barrier by 0.12 eV. Furthermore, B/P co-doping of graphene facilitated reduction both in the O_2 evolution activation barrier by 0.70 eV and the charge energy by 0.13 eV [121].

4. Oxygen Reduction and Oxygen Evolution Reaction on Bifunctional 2D Catalysts

Combining the two functionalities (OER and ORR) in a single bifunctional oxygen redox catalyst would greatly simplify the design of energy conversion devices or enhance the mobility and power-to-weight ratio. To this end, research endeavor has been devoted to exploring suitable bifunctional catalytic surface sites, or designing two distinct active sites on one material. To evaluate the activity of a bifunctional catalyst, an overvoltage between the ORR and the OER ($E_{OER@10\ mA\ cm^{-2}} - E_{ORR@-3\ mA\ cm^{-2}}$), which translates into a loss in efficiency, has been developed as a descriptor.

As is well-known, graphene doped with nitrogen appears as a promising catalyst for ORR. Nitrogen-doped graphene can also act as a bifunctional catalyst for ORR and OER, as reported by Wong et al. and other independent groups [122, 123]. DFT results suggested that the nitrogen substitution of carbon situated at the edges of armchair graphene nanoribbons resulted in low overpotentials of 0.41 and 0.45 V for the OER and ORR, respectively, which are comparable to those of Pt-containing catalysts [124]. OER and ORR catalysis can take place near the edges of the same structure but on different active sites. Recently, N/P co-doped graphene/carbon nanosheets were reported as the best bifunctional oxygen electrode catalyst hitherto [61, 125, 126]. Notably, the N and P co-doping generated synergistic effects to improve the electrocatalytic activities towards both OER and ORR. These nanosheets exhibited a potential gap of 0.71 V between the OER potential at 10 mA cm^{-2} and the ORR potential at -3 mA cm^{-2} [125, 126]. Based on thermodynamically theoretical analysis, those individual N-doped, P-doped and the N/P-co-doped nanosheets were shown to have minimum ORR overpotentials of 0.44, 0.47 and 0.47 V, respectively, while the lowest OER overpotentials for these structures were 0.41, 0.49 and 0.39 V, respectively. B-G was also shown as a bifunctional catalyst for both ORR and OER. A current density of 5 mA cm^{-2} at 0.9 V was observed when using B-G electrode for OER, outperforming the performance of commercial Pt/C (a current density of 2.4 mA cm^{-2} at 0.9 V). In the ORR, a current density of 16.32 mA cm^{-2} was attained at -0.6 V, which is approximately two times higher than that of pristine graphene (9.89 mA cm^{-2}) and is very close to the value observed for Pt/C (17.53 mA cm^{-2}) [60]. Alternatively, metal (such as Fe, Co and Ni)/heteroatom (such as N, S and O) co-doped graphene/carbon nanosheets were also studied as electrocatalysts in both ORR and OER [62-64]. As an example, Fe/N co-doped nanosheets exhibited an obviously lower overpotential 0.48 V at a current density of 10.0 mA cm^{-2} than the benchmark catalyst IrO_2 (0.46 mV). Furthermore, the addition of $NiCo_2S_4$ in N and S co-doped rGO has been shown to boost the oxidation of water into molecular oxygen. The inclusion of Ni^{2+} in Co_3S_4 brings positive effect on the overall bicatalytic activity. The mixed valence in the $NiCo_2S_4$ nanocrystals facilitates electron transfer and O_2 reversible absorption, affording a current density of 10 mA cm^{-2} at a small overpotential of ~0.47 V, which is close to the best performance of the well-investigated Co_3O_4/graphene nanosheet hybrid. Likewise, N-doped Co_9S_8 nanosheets coupling with graphene were demonstrated to exhibit highly efficient activities towards ORR (the onset potential is 0.941 V vs. RHE) and OER (the low onset potential is ~1.51 V vs. RHE) [59]. The Tafel slope of the resulting N-Co_9S_8/G (82.7 mV dec^{-1}) is significantly lower than that of commercial RuO_2/C (157.5 mV dec^{-1}).

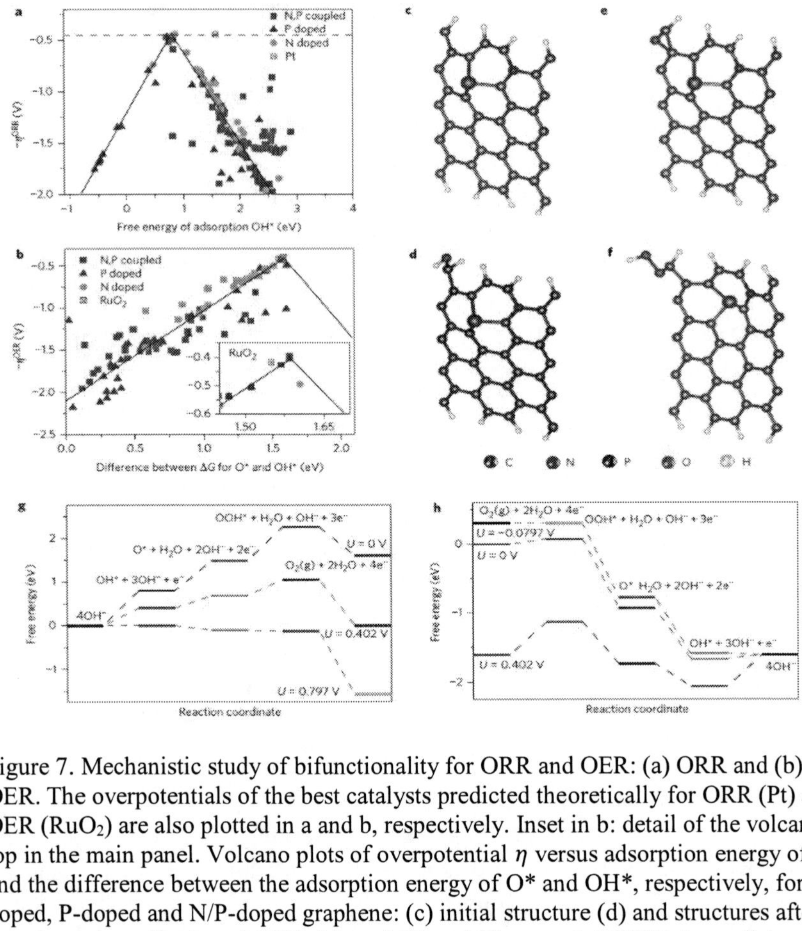

Figure 7. Mechanistic study of bifunctionality for ORR and OER: (a) ORR and (b) OER. The overpotentials of the best catalysts predicted theoretically for ORR (Pt) and OER (RuO_2) are also plotted in a and b, respectively. Inset in b: detail of the volcano top in the main panel. Volcano plots of overpotential η versus adsorption energy of O* and the difference between the adsorption energy of O* and OH*, respectively, for N-doped, P-doped and N/P-doped graphene: (c) initial structure (d) and structures after the adsorption of hydroxyl OH*, (e) oxyl O* and (f) peroxyl OOH* intermediates on N and P coupled graphene. O*, OH* and OOH* are adsorbed intermediates. (g) Schematic energy profiles for the OER pathway and (h) the ORR pathway on N/P co-doped graphene in alkaline media [110].

A more recent work performed by Qiao and co-workers showed that assembly of 2D *g*-C_3N_4 and MXene (Ti_3C_2) can accelerate oxygen electrochemistry, attaining very high activity and strong durability. A porous *g*-C_3N_4/Ti_3C_2 film showed a sharp onset potential of 1.44 V (vs. RHE) and significantly higher OER current density than that of (20 wt%) IrO_2 supported on carbon black [74]. This excellent performance was ascribed to Ti-N_x motifs acting as electroactive sites, and a hierarchically porous structure with a highly hydrophilic surface.

CONCLUSION

A number of nanosheet materials are now accessible, and an exhaustive study of how they work in catalysis is still in their infancy. The "quasi all-surface" nature of a 2D nanosheet offers unusual catalytic properties for diverse chemical processes especially for oxygen electrochemical reactions. In particular, controlling surface structure by using doping, heterostructural and alloying strategies provides new opportunities for fine tuning of active sites to enhance surface properties for catalysis. Promising results have already been attained in developing 2D materials in oxygen reduction reaction and oxygen evolution reaction competitive with noble metal catalysts, stimulating a practical goal of decreasing our dependence on rare and expensive metallic catalysts.

Chemical doping of graphene with a single heteroatom such as B, N, S, P, halogen, transition metals (e.g., Fe, Co, Ni) and multiple elements (B/N, S/N, P/N, B/P, B/N/P, N/P/S, N/Fe, N/S/Fe etc.) allows one to tune its catalytic behaviors. Nanosheets such as nickel nitride and metal oxides rich in oxygen vacancies also show potential as electrocatalysts in OER. To attain minimum free energy change during reaction on a given surface, a number of novel 2D composite materials were designed and synthesized. Integrated metal/graphene materials have provided catalysts, which were competitive against IrO_2, the state of the art catalyst for the OER. Carbon nitride/MXene (Ti_3C_2) has also been found to outweigh carbon-supported IrO_2. Heterostructures of MoS_2 and porous carbon gave ORR half-wave potentials 108 mV more positive than MoS_2 nanosheets by itself, and oxygen-doped MoS_2 nanosheets yielded a 160 mV positive shift in half-wave potential and an 8-fold enhancement in ORR activity than undoped nanosheets. We believe that the merging theoretical calculations and in-situ characterization techniques would greatly promote the molecular design and nanostructure engineering of new two dimensional catalysts toward oxygen electrocatalytic reactions.

REFERENCES

[1] Yang, Z. G. Electrochemical energy storage for green grid. *Chem Rev* 111, 3577-3613 (2011).

[2] Mai, L. Q., Tian, X. C., Xu, X., Chang, L. & Xu, L. Nanowire electrodes for electrochemical energy storage devices. *Chem Rev* 114, 11828-11862 (2014).
[3] Kang, D. et al. Electrochemical synthesis of photoelectrodes and catalysts for Use in Solar Water Splitting. *Chem Rev* 115, 12839-12887 (2015).
[4] Hisatomi, T., Kubota, J. & Domen, K. Recent advances in semiconductors for photocatalytic and photoelectrochemical water splitting. *Chem Soc Rev* 43, 7520-7535 (2014).
[5] Ferrari, A. C. Science and technology roadmap for graphene, related two-dimensional crystals, and hybrid systems. *Nanoscale* 7, 4598-4810 (2015).
[6] Coleman, J. N. Two-dimensional nanosheets produced by liquid exfoliation of layered materials. *Science* 331, 568-571 (2011).
[7] Geim, A. K. & Novoselov, K. S. The rise of graphene. *Nat Mater* 6, 183-191 (2007).
[8] Novoselov, K. S. A roadmap for graphene. *Nature* 490, 192-200 (2012).
[9] Chhowalla, M. et al. The chemistry of two-dimensional layered transition metal dichalcogenide nanosheets. *Nat Chem* 5, 263-275 (2013).
[10] Wang, H. T., Yuan, H. T., Hong, S. S., Li, Y. B. & Cui, Y. Physical and chemical tuning of two-dimensional transition metal dichalcogenides. *Chem Soc Rev* 44, 2664-2680 (2015).
[11] Chhowalla, M., Liu, Z. F. & Zhang, H. Two-dimensional transition metal dichalcogenide (TMD) nanosheets. *Chem Soc Rev* 44, 2584-2586 (2015).
[12] Sun, Z. Q. et al. Generalized self-assembly of scalable two-dimensional transition metal oxide nanosheets. *Nat Commun* 5 (2014).
[13] Wang, L. Z. & Sasaki, T. Titanium oxide nanosheets: graphene analogues with versatile functionalities. *Chem Rev* 114, 9455-9486 (2014).
[14] Hanlon, D. et al. Production of molybdenum trioxide nanosheets by liquid exfoliation and their application in high-performance supercapacitors. *Chem Mater* 26, 1751-1763 (2014).
[15] Pacchioni, G. Two-dimensional oxides: multifunctional materials for advanced technologies. *Chem-Eur J* 18, 10144-10158 (2012).
[16] Liu, L. Heteroepitaxial growth of two-dimensional hexagonal boron nitride templated by graphene edges. *Science* 343, 163-167 (2014).
[17] Liu, J., Wang, H. Q. & Antonietti, M. Graphitic carbon nitride "reloaded": emerging applications beyond (photo) catalysis. *Chem Soc Rev* 45, 2308-2326 (2016).

[18] Naguib, M., Mochalin, V. N., Barsoum, M. W. & Gogotsi, Y. 25th Anniversary article: MXenes: a new family of two-dimensional materials. *Adv Mater* 26, 992-1005 (2014).
[19] Sun, Y. F., Gao, S., Lei, F. C. & Xie, Y. Atomically-thin two-dimensional sheets for understanding active sites in catalysis. *Chem Soc Rev* 44, 623-636 (2015).
[20] Deng, D. H. et al. Catalysis with two-dimensional materials and their heterostructures. *Nat Nanotechnol* 11, 218-230 (2016).
[21] Sun, Y. F., Gao, S. & Xie, Y. Atomically-thick two-dimensional crystals: electronic structure regulation and energy device construction. *Chem Soc Rev* 43, 530-546 (2014).
[22] Lv, R. T. et al. Transition metal dichalcogenides and beyond: synthesis, properties, and applications of single- and few-layer nanosheets. *Accounts Chem Res* 48, 897-897 (2015).
[23] Yu, L., Pan, X. L., Cao, X. M., Hu, P. & Bao, X. H. Oxygen reduction reaction mechanism on nitrogen-doped graphene: A density functional theory study. *J Catal* 282, 183-190 (2011).
[24] Zhang, C. S. a. J. *PEM fuel cell electrocatalysts and catalyst layers*. 89–134 (Springer, 2008).
[25] Yeager, E. Dioxygen electrocatalysis: mechanism in relation to catalyst structure. *J Mol Catal* 38, 5-25 (1986).
[26] Yan, H. J., Xu, B., Shi, S. Q. & Ouyang, C. Y. First-principles study of the oxygen adsorption and dissociation on graphene and nitrogen doped graphene for Li-air batteries. *J Appl Phys* 112 (2012).
[27] Sheng, Z. H., Gao, H. L., Bao, W. J., Wang, F. B. & Xia, X. H. Synthesis of boron doped graphene for oxygen reduction reaction in fuel cells. *J Mater Chem* 22, 390-395 (2012).
[28] Sanetuntikul, S. S. G. J. J. Boron and phosphorous-doped graphene as metal-free electrocatalyst for oxygen reduction reaction in alkaline medium. *Rsc Adv* 5, 141-145 (2015).
[29] Jiao, Y., Zheng, Y., Jaroniec, M. & Qiao, S. Z. Origin of the electrocatalytic oxygen reduction activity of graphene-based catalysts: A roadnnap to achieve the best performance. *J Am Chem Soc* 136, 4394-4403 (2014).
[30] Sun, Z. Y. et al. High-quality functionalized few-layer graphene: facile fabrication and doping with nitrogen as a metal-free catalyst for the oxygen reduction reaction. *J Mater Chem A* , 15444-15450 (2015).

[31] Huijun Yu, L. S., Tong Bian, Run Shi, Geoffrey I. N. Waterhouse, Yufei Zhao, Chao Zhou, Li-Zhu Wu, Chen-Ho Tung, Tierui Zhang. Nitrogen-doped porous carbon nanosheets templated from g-C3N4 as metal-free electrocatalysts for efficient oxygen reduction reaction. *Adv. Mater* 28, 5080-5086 (2016).

[32] Zhang, C. Z., Mahmood, N., Yin, H., Liu, F. & Hou, Y. L. Synthesis of phosphorus-doped graphene and its multifunctional applications for oxygen reduction reaction and lithium ion batteries. *Adv Mater* 25, 4932-4937 (2013).

[33] Zi-Wu Liu, F. P., Hong-Juan Wang, Hao Yu, Wen-Xu Zheng, and Jian Yang. Phosphorus-doped graphite layers with high electrocatalytic activity for the O_2 reduction in an alkaline medium. *Angewandte Chemie International Edition* 50, 3257-3261 (2011).

[34] Park, J. E. et al. Sulfur-doped graphene as a potential alternative metal-free electrocatalyst and Pt-catalyst supporting material for oxygen reduction reaction. *Phys Chem Chem Phys* 16, 103-109, (2014).

[35] Yang Z, Y. Z., Li G, Fang G, Nie H, Liu Z, Zhou X, Chen X, Huang S. Sulfur-doped graphene as an efficient metal-free cathode catalyst for oxygen reduction. *Acs Nano* 6, 205-211 (2012).

[36] Jeon, I. Y. et al. Facile, scalable synthesis of edge-halogenated graphene nanoplatelets as efficient metal-free eletrocatalysts for oxygen reduction reaction. *Sci Rep-Uk* 3, 1810-1816 (2013).

[37] Jin, J. T. et al. Catalyst-free synthesis of crumpled boron and nitrogen co-doped graphite layers with tunable bond structure for oxygen reduction reaction. *Acs Nano* 8, 3313-3321 (2014).

[38] Choi CH, P. S., Woo SI. Binary and ternary doping of nitrogen, boron, and phosphorus into carbon for enhancing electrochemical oxygen reduction activity. *Acs Nano* 6, 7084-7091 (2012).

[39] Xu, J. X., Dong, G. F., Jin, C. H., Huang, M. H. & Guan, L. H. Sulfur and nitrogen co-doped, few-layered graphene oxide as a highly efficient electrocatalyst for the oxygen-reduction reaction. *Chemsuschem* 6, 493-499 (2013).

[40] Ai, W. et al. Nitrogen and sulfur codoped graphene: Multifunctional electrode materials for high-performance Li-ion batteries and oxygen reduction reaction. *Adv Mater* 26, 6186-6192 (2014).

[41] Akhter, T. et al. Self-assembled N/S codoped flexible graphene paper for high performance energy storage and oxygen reduction reaction. *Acs Appl Mater Inter* 8, 2078-2087 (2016).

[42] Dou, S. et al. N-, P- and S-tridoped graphene as metal-free electrocatalyst for oxygen reduction reaction. *J Electroanal Chem* 753, 21-27 (2015).

[43] Yang, S. B., Feng, X. L., Wang, X. C. & Mullen, K. Graphene-based carbon nitride nanosheets as efficient metal-free electrocatalysts for oxygen reduction reactions. *Angew Chem Int Edit* 50, 5339-5343 (2011).

[44] Funatsu, A. et al. Synthesis of monolayer platinum nanosheets. *Chem Commun* 50, 8503-8506 (2014).

[45] Zhong, X. et al. Pt@Au nanorods uniformly decorated on pyridyne cycloaddition graphene as a highly effective electrocatalyst for oxygen reduction. *Acs Appl Mater Inter* 6, 13448-13454 (2014).

[46] Men, B. et al. Highly dispersed Ag-functionalized graphene electrocatalyst for oxygen reduction reaction in energy-saving electrolysis of sodium carbonate. *Ind Eng Chem Res* 54, 7415-7422, (2015).

[47] Shi, C. et al. Synthesis of layered MnO_2 nanosheets for enhanced oxygen reduction reaction catalytic activity. *Electrochim Acta* 132, 239-243 (2014).

[48] Liu, B. et al. Unconventional structural and morphological transitions of nanosheets, nanoflakes and nanorods of AuNP@MnO2. *J Mater Chem A* 4, 6447-6455 (2016).

[49] Kai Yuan, X. Z., Haiyan Fu, Gunther Brunklaus, Michael Forster, Yiwang Chen, Xinliang Feng, Ullrich Scherf. Two-dimensional core-shelled porous hybrids as highly efficient catalysts for the oxygen reduction reaction. *Angew Chem Int Edit* 128, 6972-6977 (2016).

[50] Huang, H., Feng, X., Du, C. C., Wu, S. Y. & Song, W. B. Incorporated oxygen in MoS2 ultrathin nanosheets for efficient ORR catalysis. *J Mater Chem A* 3, 16050-16056 (2015).

[51] Huang, J. H. et al. CoOOH Nanosheets with High Mass Activity for Water Oxidation. *Angew Chem Int Edit* 54, 8722-8727 (2015).

[52] Sun, Y. F. et al. Atomically-thin non-layered cobalt oxide porous sheets for highly efficient oxygen-evolving electrocatalysts. *Chem Sci* 5, 3976-3982 (2014).

[53] Han, X. T. et al. Mass and charge transfer coenhanced oxygen evolution behaviors in CoFe-layered double hydroxide assembled on graphene. *Adv Mater Interfaces* 3 (2016).

[54] Tang, D. et al. A high-performance reduced graphene oxide/ZnCo layered double hydroxide electrocatalyst for efficient water oxidation. *Dalton T* 43, 15119-15125 (2014).

[55] Ma, W. et al. A superlattice of alternately stacked Ni-Fe hydroxide nanosheets and graphene for efficient splitting of water. *Acs Nano* 9, 1977-1984 (2015).

[56] Feng, L. L. et al. High-index faceted Ni3S2 nanosheet arrays as highly active and ultrastable electrocatalysts for water splitting. *J Am Chem Soc* 137, 14023-14026 (2015).

[57] Fan, K. et al. Nickel-vanadium monolayer double hydroxide for efficient electrochemical water oxidation. *Nat Commun* 7, 11981-11990 (2016).

[58] Chen, S., Duan, J. J., Ran, J. R., Jaroniec, M. & Qiao, S. Z. N-doped graphene film-confined nickel nanoparticles as a highly efficient three-dimensional oxygen evolution electrocatalyst. *Energ Environ Sci* 6, 3693-3699 (2013).

[59] Dou, S., Tao, L., Huo, J., Wang, S. Y. & Dai, L. M. Etched and doped Co_9S_8/graphene hybrid for oxygen electrocatalysis. *Energ Environ Sci* 9, 1320-1326 (2016).

[60] Vineesh, T. V. et al. Bifunctional electrocatalytic activity of boron-doped graphene derived from boron carbide. *Adv Energy Mater* 5 (2015).

[61] Li, R., Wei, Z. D. & Gou, X. L. Nitrogen and phosphorus dual-doped graphene/carbon nanosheets as bifunctional electrocatalysts for oxygen reduction and evolution. *Acs Catal* 5, 4133-4142 (2015).

[62] Liang, Y. Y. et al. Co_3O_4 nanocrystals on graphene as a synergistic catalyst for oxygen reduction reaction. *Nat Mater* 10, 780-786, (2011).

[63] Liu, Q., Jin, J. T. & Zhang, J. Y. $NiCO_2S_4$@graphene as a bifunctional electrocatalyst for oxygen reduction and evolution reactions. *Acs Appl Mater Inter* 5, 5002-5008 (2013).

[64] Lee, D. U., Kim, B. J. & Chen, Z. W. One-pot synthesis of a mesoporous $NiCo_2O_4$ nanoplatelet and graphene hybrid and its oxygen reduction and evolution activities as an efficient bi-functional electrocatalyst. *J Mater Chem A* 1, 4754-4762 (2013).

[65] Guo, D. H. et al. Active sites of nitrogen-doped carbon materials for oxygen reduction reaction clarified using model catalysts. *Science* 351, 361-365, (2016).

[66] Qu, L. T., Liu, Y., Baek, J. B. & Dai, L. M. Nitrogen-doped graphene as efficient metal-free electrocatalyst for oxygen reduction in fuel cells. *Acs Nano* 4, 1321-1326 (2010).

[67] Lai, L. F. et al. Exploration of the active center structure of nitrogen-doped graphene-based catalysts for oxygen reduction reaction. *Energ Environ Sci* 5, 7936-7942, (2012).

[68] Ma, F. X., Wang, J., Wang, F. B. & Xia, X. H. The room temperature electrochemical synthesis of N-doped graphene and its electrocatalytic activity for oxygen reduction. *Chem Commun* 51, 1198-1201 (2015).

[69] Wu, Z. S. et al. Three-dimensional nitrogen and boron Co-doped graphene for high-performance all-solid-state supercapacitors. *Adv Mater* 24, 5130-5135 (2012).

[70] Li, X. et al. Boron doping of graphene for graphene-silicon p-n junction solar cells. *Adv Energy Mater* 2, 425-429 (2012).

[71] Dewhurst, R. D., Claessen, R. & Braunschweig, H. Two-dimensional, but not flat: an all-boron graphene with a corrugated structure. *Angew Chem Int Edit* 55, 4866-4868 (2016).

[72] Zhang, L. P., Niu, J. B., Li, M. T. & Xia, Z. H. Catalytic mechanisms of sulfur-doped graphene as efficient oxygen reduction reaction catalysts for fuel cells. *J Phys Chem C* 118, 3545-3553, (2014).

[73] Jeon, I. Y. et al. Edge-selectively sulfurized graphene nanoplatelets as efficient metal-free electrocatalysts for oxygen reduction reaction: The electron spin effect. *Adv Mater* 25, 6138-6145, (2013).

[74] Ma, T. Y., Cao, J. L., Jaroniec, M. & Qiao, S. Z. Interacting carbon nitride and titanium carbide nanosheets for high-performance oxygen evolution. *Angew Chem Int Edit* 55, 1138-1142 (2016).

[75] Shubin Yang, L. Z., Kun Tang, Xinliang Feng,Joachim Maier,Klaus Müllen. Efficient synthesis of heteroatom (N or S)-doped graphene based on ultrathin graphene oxide-porous silica sheets for oxygen reduction rReactions. *Adv. Funct. Mater* 22, 3634-3640 (2012).

[76] Liu, Z. W. et al. Phosphorus-doped graphite layers with high electrocatalytic activity for the oxygen reduction in an alkaline medium. *Angew Chem Int Edit* 50, 3257-3261 (2011).

[77] Kakaei, K. & Balavandi, A. Synthesis of halogen-doped reduced graphene oxide nanosheets as highly efficient metal-free electrocatalyst for oxygen reduction reaction. *J Colloid Interf Sci* 463, 46-54 (2016).

[78] Norskov, J. K. et al. Origin of the overpotential for oxygen reduction at a fuel-cell cathode. *J Phys Chem B* 108, 17886-17892, (2004).

[79] Hou, S. C. et al. Nitrogen-doped graphene for dye-sensitized solar cells and the role of nitrogen states in triiodide reduction. *Energ Environ Sci* 6, 3356-3362 (2013).

[80] Tang, Y. B. et al. Tunable band gaps and p-Type transport properties of boron-doped graphenes by controllable ion doping using reactive microwave plasma. *Acs Nano* 6, 1970-1978 (2012).

[81] Agnoli, S. & Favaro, M. Doping graphene with boron: a review of synthesis methods, physicochemical characterization, and emerging applications. *J Mater Chem A* 4, 5002-5025 (2016).

[82] Cloke, R. R. et al. Site-specific substitutional boron doping of semiconducting armchair graphene nanoribbons. *J Am Chem Soc* 137, 8872-8875 (2015).

[83] Wang, X. W. et al. Heteroatom-doped graphene materials: syntheses, properties and applications. *Chem Soc Rev* 43, 7067-7098 (2014).

[84] Ricardo Faccio, L. F.W., Helena Pardo, Cecilia Goyenola, Oscar N. Ventura, Álvaro W. Mombrú Electronic and structural distortions in graphene induced by carbon vacancies and boron doping. *J. Phys. Chem. C* 114, 18961-18971 (2010).

[85] Kawai, S. et al. Atomically controlled substitutional boron-doping of graphene nanoribbons. *Nat Commun* 6, 9098-9104 (2015).

[86] Chia, X. Y., Eng, A. Y. S., Ambrosi, A., Tan, S. M. & Pumera, M. Electrochemistry of nanostructured layered transition-metal dichalcogenides. *Chem Rev* 115, 11941-11966 (2015).

[87] Zhao L, H. R., Rim KT, Schiros T, Kim KS, Zhou H, Gutiérrez C, Chockalingam SP, Arguello CJ, Pálová L, Nordlund D, Hybertsen MS, Reichman DR, Heinz TF Kim P, Pinczuk A, Flynn GW, Pasupathy AN. Visualizing individual nitrogen dopants in monolayer graphene. *Science* 333, 999-10003 (2011).

[88] Fazio, G., Ferrighi, L. & Di Valentin, C. Boron-doped graphene as active electrocatalyst for oxygen reduction reaction at a fuel-cell cathode. *J Catal* 318, 203-210, (2014).

[89] Wang, X. N-doping of graphene through electrothermal reactions with ammonia. *Science* 329, 1467-1467 (2009).

[90] Li, X. L. et al. Simultaneous nitrogen doping and reduction of graphene oxide. *J Am Chem Soc* 131, 15939-15944 (2009).

[91] L. S. Panchakarla, K. S. S., S. K. Saha, Achutharao Govindaraj, H. R. Krishnamurthy, U. V. Waghmare, C. N. R. Rao. Synthesis, structure and properties of boron and nitrogen doped graphene. *Adv Mater* 21, 4726-4730 (2009).

[92] Chaohua Zhang, L. F., Nan Liu, Minhao Liu, Yayu Wang, Zhongfan Liu. Synthesis of nitrogen-doped graphene using embedded carbon and nitrogen sources. *Adv Mater* 23, 1020-1024 (2011).

[93] Joucken, F. et al. Charge transfer and electronic doping in nitrogen-doped graphene. *Sci Rep-Uk* 5, 14564-14574 (2015).

[94] Wang, H. M. et al. Phosphorus-doped graphene and (8,0) carbon nanotube: Structural, electronic, magnetic properties, and chemical reactivity. *Appl Surf Sci* 273, 302-309 (2013).

[95] Wu, J. et al. Synthesis and electrocatalytic activity of phosphorus-doped carbon xerogel for oxygen reduction. *Electrochim Acta* 127, 53-60 (2014).

[96] Zhang, X. L. et al. The mechanisms of oxygen reduction reaction on phosphorus doped graphene: A first-principles study. *J Power Sources* 276, 222-229, (2015).

[97] Denis, P. A., Faccio, R. & Mombru, A. W. Is it possible to dope single-walled carbon nanotubes and graphene with sulfur? *Chemphyschem* 10, 715-722 (2009).

[98] Zheng, Y., Jiao, Y., Ge, L., Jaroniec, M. & Qiao, S. Z. Two-step boron and nitrogen doping in graphene for enhanced synergistic catalysis. *Angew Chem Int Edit* 52, 3110-3116 (2013).

[99] Sen, D., Thapa, R. & Chattopadhyay, K. K. Rules of boron-nitrogen doping in defect graphene sheets: A first-principles investigation of band-gap tuning and oxygen reduction reaction catalysis capabilities. *Chemphyschem* 15, 2542-2549, (2014).

[100] Ma, R. G. et al. Ionic liquid-assisted synthesis of dual-doped graphene as efficient electrocatalysts for oxygen reduction. *Carbon* 102, 58-65 (2016).

[101] Zhang, H. H. et al. Bioinspired synthesis of nitrogen/sulfur co-doped graphene as an efficient electrocatalyst for oxygen reduction reaction. *J Power Sources* 279, 252-258, (2015).

[102] Hualin Lin, L. C., Xinjing Wang, Zhaoquan Yao, Fan Liu, Yani Ai, Xiaodong Zhuang and Sheng Han. Boron, nitrogen, and phosphorous ternary doped graphene aerogel with hierarchically porous structures as highly efficient electrocatalysts for oxygen reduction reaction. *New J. Chem.* 40, 6022--6029 (2016).

[103] Yu,et al. Enhanced electrocatalytic activity due to additional phosphorous doping in nitrogen and sulfur-doped graphene: A comprehensive study. *Carbon* 78, 257-267 (2014).

[104] Ci, L. et al. Atomic layers of hybridized boron nitride and graphene domains. *Nat Mater* 9, 430-435 (2010).

[105] Liu, Z. et al. In-plane heterostructures of graphene and hexagonal boron nitride with controlled domain sizes. *Nat Nanotechnol* 8, 119-124 (2013).

[106] Zhao, Y. et al. Can boron and nitrogen Co-doping improve oxygen reduction reaction activity of carbon nanotubes? *J Am Chem Soc* 135, 1201-1204 (2013).

[107] Choi, C. H., Chung, M. W., Kwon, H. C., Park, S. H. & Woo, S. I. B, N- and P, N-doped graphene as highly active catalysts for oxygen reduction reactions in acidic media. *J Mater Chem A* 1, 3694-3699 (2013).
[108] Kamiya, K., Hashimoto, K. & Nakanishi, S. Instantaneous one-pot synthesis of Fe-N-modified graphene as an efficient electrocatalyst for the oxygen reduction reaction in acidic solutions. *Chem Commun* 48, 10213-10215 (2012).
[109] Elumalai, G., Noguchi, H. & Uosaki, K. Electrocatalytic activity of various types of *h*-BN for the oxygen reduction reaction. *Phys Chem Chem Phys* 16, 13755-13761 (2014).
[110] Zhao, J. X. & Chen, Z. F. Carbon-doped boron nitride nanosheet: An efficient metal-free electrocatalyst for the oxygen reduction reaction. *J Phys Chem C* 119, 26348-26354 (2015).
[111] Shin, D., Thapa, R. & Park, N. An oxygen reduction catalytic process through superoxo adsorption states on n-type doped *h*-BN: A first-principles study. *Curr Appl Phys* 15, 727-732 (2015).
[112] Duan, J. J., Chen, S., Jaroniec, M. & Qiao, S. Z. Porous C_3N_4 nanolayers@N-graphene films as catalyst electrodes for highly efficient hydrogen evolution. *Acs Nano* 9, 931-940 (2015).
[113] Zheng, Y. et al. Nanoporous graphitic-C_3N_4@carbon metal-free electrocatalysts for highly efficient oxygen reduction. *J Am Chem Soc* 133, 20116-20119 (2011).
[114] He, W. et al. An efficient reduction route for the production of Pd-Pt nanoparticles anchored on graphene nanosheets for use as durable oxygen reduction electrocatalysts. *Carbon* 50, 265-274 (2012).
[115] Xu, K. et al. Metallic nickel nitride nanosheets realizing enhanced electrochemical water oxidation. *J Am Chem Soc* 137, 4119-4125 (2015).
[116] Song, F. & Hu, X. L. Exfoliation of layered double hydroxides for enhanced oxygen evolution catalysis. *Nat Commun* 5 (2014).
[117] Qiao, S. C. J. D. J. R. S. Pape-based N-doped carbon films for enhanced oxygen evolution electrocatalysis. *Adv Sci* 2, 1-2 (2015).
[118] Mirzakulova, E. et al. Electrode-assisted catalytic water oxidation by a flavin derivative. *Nat Chem* 4, 794-801 (2012).
[119] Zhao, Y., Nakamura, R., Kamiya, K., Nakanishi, S. & Hashimoto, K. Nitrogen-doped carbon nanomaterials as non-metal electrocatalysts for water oxidation. *Nat Commun* 4 (2013).

[120] Tan, X., Tahini, H. A., Seal, P. & Smith, S. C. First-principle framework for total charging energies in electrocatalytic materials and charge-responsive molecular binding at gas-surface interfaces. *Acs Appl Mater Inter* 8, 10897-10903 (2016).

[121] Ren, X. D. et al. The doping effect on the catalytic activity of graphene for oxygen evolution reaction in a lithium-air battery: a first-principles study. *Phys Chem Chem Phys* 17, 14605-14612 (2015).

[122] Lin, Z. Y., Waller, G. H., Liu, Y., Liu, M. L. & Wong, C. P. Simple preparation of nanoporous few-layer nitrogen-doped graphene for use as an efficient electrocatalyst for oxygen reduction and oxygen evolution reactions. *Carbon* 53, 130-136 (2013).

[123] Wang, L., Yin, F. X. & Yao, C. X. N-doped graphene as a bifunctional electrocatalyst for oxygen reduction and oxygen evolution reactions in an alkaline electrolyte. *Int J Hydrogen Energ* 39, 15913-15919 (2014).

[124] Li, M. T., Zhang, L. P., Xu, Q., Niu, J. B. & Xia, Z. H. N-doped graphene as catalysts for oxygen reduction and oxygen evolution reactions: Theoretical considerations. *J Catal* 314, 66-72 (2014).

[125] Hong Bin Yang, J. M., Sung-Fu Hung, Jiazang Chen, Hua Bing Tao, Xizu Wang, Liping Zhang, Rong Chen, Jiajian Gao, Hao Ming Chen, Liming Dai, and Bin Liu1. Identification of catalytic sites for oxygen reduction and oxygen evolution in N-doped graphene materials: Development of highly efficient metal-free bifunctional electrocatalyst. *Science Advances* 2, e1501122 (2016).

[126] Zhang, J. T., Zhao, Z. H., Xia, Z. H. & Dai, L. M. A metal-free bifunctional electrocatalyst for oxygen reduction and oxygen evolution reactions. *Nat Nanotechnol* 10, 444-452 (2015).

Index

2

2D materials, vii, ix, 133, 135, 154, 159

A

absorption spectra, 111
acetaminophen, 19, 37, 54
acetone, 90, 109
acetonitrile, 61
acetylcholinesterase, 81
acid, 10, 15, 16, 22, 23, 28, 32, 37, 55, 81, 90, 91
acidic, 15, 17, 18, 135, 140, 143, 144, 148, 150, 154, 168
acrylate, 90, 91, 110
acrylonitrile, 9
activated carbon, 15, 23
activation state, 153
active centers, 145, 156
active compound, 34, 36, 38, 40
active site, 135, 141, 144, 146, 148, 149, 154, 156, 157, 159, 161
adenine, 55, 74
adenosine, 47, 48, 49
adenosine triphosphate, 47, 48
adhesion, 87
adsorption, 12, 13, 15, 17, 18, 19, 21, 23, 24, 137, 142, 144, 145, 146, 147, 151, 153, 154, 158, 161, 168
advanced oxidation processes, 23, 39
AFM, ix, 85, 87, 89, 90, 92, 93, 94, 95, 96, 97, 98, 99, 100, 102, 103, 104, 105, 110, 111, 112, 113, 115, 116, 117, 123
aggregation, 106, 109
alkaline media, 142, 151, 158
ALS, 46
alternate compression–relaxation method, ix, 86
alternative energy, 24
amino acids, 36
amino groups, 11
ammonium, 45, 54, 73, 92, 152
angstroms, 87
annealing, 143, 145, 146, 147, 152
antibiotic resistance, 14
aqueous solutions, 36, 71
argon, 146, 147
aromatic polyamide, viii, 85, 87, 89, 90, 92, 93, 94, 101, 105, 106
ascorbic acid, 54, 56, 73
assessment, 33, 36
asymmetry, 148
atomic force, ix, 85, 87
atoms, 43, 55, 135, 142, 143, 145, 146, 147, 152, 154

ATP, 47, 70
Au nanoparticles, 72, 75

B

bacteria, vii, 1, 5, 16, 71
band gap, 61, 142, 147, 152, 165
batteries, 125, 134, 135, 153, 161, 162
benchmarking, 155
binding energy, 153
biochemistry, 86
biocompatibility, 42
biomolecules, 53, 61, 69, 72, 78
biopolymers, 83
biosensors, 69, 72, 77, 78
bisphenol, 54, 72, 73
bonding, 6, 25, 143, 146, 147
bonds, 135, 142, 143, 145, 146
bottom-up, 86, 123
BPA, 23

C

CAF, 23
caffeine, 20, 23
calcium, 21, 37
calcium carbonate, 21
calibration, 64
cancer, 61, 77
candidates, viii, 41, 43, 154
carbamazepine, 14, 15, 17, 18, 22, 23, 24, 35
carbon, vii, viii, 2, 6, 13, 25, 28, 29, 31, 44, 54, 56, 57, 65, 72, 73, 75, 76, 81, 82, 90, 134, 137, 139, 142, 143, 145, 146, 147, 149, 151, 155, 157, 158, 159, 160, 162, 163, 164, 165, 166, 167, 168
carbon atoms, 2, 134, 143, 145, 146, 147, 156
carbon film, 168
carbon materials, 149, 164
carbon nanotubes, 28, 31, 73, 155, 167
carboxyl, 8
carboxylic acid, 3, 10, 91
carcinoembryonic antigen, 49
case study(ies), 3, 82
catalysis, viii, 41, 42, 43, 135, 154, 157, 159, 160, 161, 163, 167, 168
catalyst, 46, 68, 137, 141, 142, 143, 149, 150, 151, 153, 154, 156, 159, 161, 162, 164, 168
catalytic activity, 42, 44, 45, 55, 59, 68, 77, 81, 146, 153, 154, 156, 163, 169
catalytic properties, 150, 159
cation, 92
C-C, 142, 145
cell membranes, 86
cellulose, 8, 13, 39
ceramic, vii, viii, 2, 3, 6, 14, 23, 26
cerium, 44
chalcogenides, 150
challenges, viii, ix, 2, 41, 63, 125, 133
charge density, 3, 20, 142, 146
chemical, vii, ix, 1, 3, 10, 13, 15, 17, 28, 34, 35, 36, 42, 43, 54, 62, 78, 79, 80, 115, 133, 134, 135, 141, 159, 160, 167
chemical interaction, 17
chemical properties, ix, 36, 42, 43, 133, 135
chemical reactivity, 167
chemical vapor deposition, 62, 78
chemiluminescence, 44, 52, 59, 72, 77
chemisorption, 142
chemometrics, 46, 52, 80, 82, 83
China, 41, 44, 64, 74, 80, 133
chitosan, 9, 11, 29, 31, 54, 57, 72, 127
chloride anion, 11
chlorination, 15
chlorine, 8, 9, 15
chloroform, 62, 92, 103, 104, 110, 122
cholesterol, 46, 59, 68, 77
chromatography, 43, 110
clean energy, 134, 153
cleaning, vii, 1, 4, 9, 13, 35, 38
cleavage, 50, 142
clusters, 82
C-N, 145, 148

coatings, 87
cobalt, 163
comb copolymer, ix, 86, 90, 92, 98, 99, 100, 101, 102, 105, 106, 109, 123
commercial, 13, 35, 92, 145, 150, 151, 154, 157
community, viii, 41
complementary DNA, 51
composites, 32, 54, 66, 72, 73, 75, 76, 134, 144, 154
composition, 8, 9, 17
compounds, vii, 1, 4, 8, 13, 14, 15, 16, 17, 18, 20, 21, 23, 33, 35, 36, 55, 90, 102, 154
compression, ix, 86, 92, 93, 95, 101, 109, 110, 112, 113, 114, 115, 116, 117, 124
condensation, 90
conduction, 25, 142, 143
conductivity, 25, 42, 56, 59, 63, 149, 154
configuration, 142, 143, 144, 146, 148
conjugation, 144
consensus, 148
construction, 47, 64, 87, 102, 105, 161
consumption, 134
contamination, 14, 24
coordination, 43, 55, 135, 154
copolymer, ix, 14, 31, 86, 90, 98, 99, 100, 101, 105, 109, 110, 111, 112, 114, 115, 116, 118, 119, 120, 121, 122, 123, 124
copolymerization, 90, 99, 109
copolymers, 10, 90, 91, 92, 98, 99, 100, 101, 102, 106, 109, 110, 116, 118
correlation, 17
cosmetics, 87
cost, 4, 25, 26, 43, 44, 134, 140, 150, 156
crystal structure, 42
crystalline, 79, 86
crystallinity, 119
crystals, ix, 86, 87, 106, 109, 119, 120, 124, 134, 160, 161
CVD, 61, 150
cycles, 35
cysteine, 50, 55, 73, 82
cytochrome, 70

D

decomposition, 64, 156
deconvolution, 109
defects, 3, 11, 18, 111, 113, 115, 135
deformation, 63
degradation, 9, 12, 13, 14, 15, 23, 31, 39, 150
dehydration, 12
density functional theory, 136, 161
Department of Education, 64
deposition, 6, 9, 10, 13, 21
derivatives, vii, viii, ix, 2, 4, 59, 77, 105, 106, 133, 135, 141, 142
desalination, 2, 4, 5, 7, 8, 9, 11, 16, 20, 26, 28, 29, 30, 31, 33, 37, 38, 39
desorption, 47
detachment, 62
detection, 43, 44, 46, 47, 48, 49, 50, 51, 52, 53, 57, 58, 59, 60, 63, 64, 68, 69, 70, 71, 72, 73, 74, 75, 76, 77, 78, 79, 81, 82
DFT, 136, 142, 145, 149, 157
diallyldimethylammonium chloride, 54, 73, 151
diclofenac, 14, 15, 17, 21, 23, 33, 35
diffraction, 98
diffusion, vii, 1, 9, 12, 17, 18, 20, 22, 25, 147
diffusion process, 18
discrimination, 50, 59
disinfection, 8, 15
dispersion, 4, 10, 91, 92
dissociation, 156, 161
distillation, vii, 1, 3, 12, 31
distilled water, 92, 103, 104, 110
distortions, 166
distribution, 17, 110, 113, 115, 116, 117
DNA, 47, 48, 51, 53, 57, 58, 59, 62, 69, 70, 71, 75, 76, 77, 79, 81
dopamine, 54, 56, 73, 74

dopants, 142, 144, 145, 148, 166
doping, 46, 51, 61, 77, 135, 141, 142, 143, 144, 145, 147, 148, 151, 156, 157, 159, 161, 162, 165, 166, 167, 169
double bonds, 118
drinking water, 4, 14, 15, 16, 21, 24, 31, 34, 35, 39
drug delivery, 42, 65
durability, 93, 140, 147, 151, 158
dyes, 5, 11, 47, 69

E

effluent, 14, 15, 33, 39, 40
electric field, 23, 59, 63
electrical conductivity, 42, 53, 141, 142, 152
electrical properties, 37, 142
electrocatalysis, 74, 134, 161, 164, 168
electrochemistry, 72, 135, 158
electrode surface, 134, 153
electrodeposition, 54
electrodes, 24, 74, 141, 147, 156, 160, 168
electrolysis, 163
electrolyte, 24, 149, 152, 155, 169
electromigration, 20
electron, 13, 51, 54, 56, 61, 86, 101, 123, 134, 135, 136, 137, 141, 142, 143, 145, 146, 147, 150, 151, 152, 153, 154, 155, 157, 165
electronic structure, ix, 133, 135, 141, 147, 152, 161
electrospinning, 6
elucidation, 108
emission, ix, 46, 52, 64, 86, 106, 120, 122, 124
endocrine, 14, 32, 35, 36
endonuclease, 49
energy, viii, 7, 8, 9, 11, 16, 24, 41, 42, 43, 46, 51, 53, 65, 66, 69, 72, 120, 134, 142, 144, 147, 149, 151, 154, 156, 158, 159, 160, 161, 162, 163
energy consumption, 8, 11
energy emission, 120
energy transfer, 46, 51, 53, 69, 72
engineering, 159
environment, 6, 33, 44, 51, 59, 63, 145
Environmental Protection Agency, 27
enzyme, 53, 55, 67, 75, 81
EPA, 15, 27
epoxy groups, 11
equilibrium, 20, 144
ethylene, 31, 130
ethylene oxide, 130
evaporation, 92, 110
evidence, 111, 113
evolution, vii, ix, 67, 133, 134, 148, 153, 156, 159, 163, 164, 165, 168, 169
excitation, 46, 52, 64, 77
exclusion, 3, 9, 16, 17, 19, 22, 26, 37
exonuclease, 49, 76
experimental condition, 142
exposure, 9, 14, 33, 61

F

Fabrication, v, viii, 2, 3, 5, 9, 22, 26, 28, 29, 30, 58, 59, 60, 75, 76, 77, 81, 85, 86, 87, 95, 126, 130, 131, 146, 150, 161
factor analysis, 52, 64
Fermi level, 141, 142
ferromagnetic, 67
ferrous ion, 71, 81
fiber, 5, 31
film formation, 113
films, 3, 10, 11, 61, 86, 87, 92, 93, 98, 99, 104, 106, 110, 111, 168
filters, 13, 107
filtration, 6, 10, 12, 13, 15, 16, 21, 26, 28, 39
fine tuning, 159
Finland, 33
flatness, 87
flavonoids, 82
flexibility, 42
fluctuations, 62
fluid, 18, 87

fluorescence, ix, 44, 46, 47, 48, 49, 50, 51, 52, 69, 70, 71, 81, 82, 86, 109, 111, 120, 124
food, 7, 43, 81
food additive, 81
food production, 7
food safety, 43
force, 93, 111
formation, ix, 18, 22, 86, 87, 90, 93, 94, 95, 97, 98, 100, 101, 102, 105, 106, 108, 109, 110, 111, 113, 115, 116, 119, 122, 124, 135, 143, 146, 148, 149
forward osmosis, vii, 1, 3, 7, 30
fouling, vii, 1, 2, 3, 8, 10, 11, 12, 13, 19, 21, 22, 23, 27, 28, 34, 35, 38, 39
free energy, 20, 22, 142, 149, 153, 159
freshwater, 33
friction, 2, 87
fuel cell, vii, 1, 2, 24, 26, 40, 134, 135, 136, 150, 161, 164, 165
fuel cells, vii, 1, 2, 24, 26, 40, 134, 135, 150, 161, 164, 165
functionalization, 3, 8, 26, 42, 108
funding, 27
fusion, 115

G

gas sensors, 78, 79
GCE, 54, 57, 59
gel, 90, 109
gel permeation chromatography, 90
glucose, 44, 53, 57, 68, 72, 73
glucose oxidase, 44, 53, 57, 72
glutamate, 127
glycol, 11
gold nanoparticles, 44, 69, 72, 75, 76, 82
GPC, 90
graphene, v, vii, viii, ix, 1, 2, 4, 8, 9, 10, 11, 13, 16, 23, 26, 27, 28, 29, 30, 31, 32, 39, 40, 42, 43, 44, 51, 53, 54, 56, 57, 58, 60, 64, 65, 66, 67, 69, 71, 72, 73, 74, 75, 76, 78, 79, 81, 82, 83, 125, 133, 134, 135, 137, 138, 139, 140, 141, 142, 143, 144, 145, 146, 147, 148, 149, 150, 151, 152, 155, 156, 158, 159, 160, 161, 162, 163, 164, 165, 166, 167, 168, 169
graphene oxide, vii, viii, 2, 8, 9, 11, 13, 26, 27, 28, 29, 30, 31, 32, 39, 40, 55, 60, 65, 81, 146, 147, 148, 151, 162, 163, 165, 166
graphene sheet, 2, 8, 167
graphite, 4, 86, 93, 111, 134, 136, 143, 144, 145, 147, 148, 156, 162, 165
groundwater, 34
growth, 33, 55, 57, 75, 76, 160
guanine, 55, 74
guidelines, 15
Guinea, 65, 125

H

halogen, 142, 147, 159, 165
hazardous organic pollutants, 14
health, 14, 32
health effects, 14
height, viii, 85, 96, 97, 101, 102, 105, 106, 113, 115, 123, 124
helium, 27
hemoglobin, 55, 74
HO-2, 136
hormone levels, 14
hormones, 15, 20
human, 32, 33, 48, 82
humidity, 61, 79
hybrid, 7, 8, 12, 27, 28, 29, 55, 64, 72, 74, 83, 151, 156, 157, 160, 164
hybridization, 47, 49, 57, 62, 70, 71, 76, 79, 87, 142, 145
hydrocarbons, 120
hydrogen, 6, 25, 44, 61, 67, 68, 73, 78, 82, 134, 168
hydrogen peroxide, 44, 68, 73, 82
hydrogenation, 149
hydrolysis, 50
hydrophilic materials, 12
hydrophilicity, viii, 2, 3, 7, 8, 9, 10, 11, 13, 14, 16, 23, 26, 28, 106

hydrophobicity, 16, 17, 21, 36, 61
hydroquinone, 54, 72
hydroxide, 154, 163, 164
hydroxyl, 3, 91, 158
hydroxyl groups, 91
hysteresis, 113
hysteresis loop, 113

I

ibuprofen, 17, 18, 20, 22, 23, 24, 35
ideal, 2, 9, 10, 91, 110
identification, 38
images, 19, 27, 60, 87, 89, 93, 94, 95, 96, 97, 98, 99, 100, 102, 103, 104, 112, 113, 115, 116, 117, 144
immobilization, 6, 12, 32, 51, 55, 57, 74
immunoglobulin, 57, 75
improvements, 9, 25
in transition, 66
induction, 108
industry(ies), vii, 1, 2, 3, 4, 34, 42, 136
in-plane XRD, 93, 98, 119, 120
integration, 24
interface, viii, 51, 85, 92, 93, 96, 97, 101, 102, 103, 104, 105, 108, 109, 110, 111, 113, 117, 118
interfacial polymerization, 8, 9, 10
interference, 87
intermolecular interactions, 105
iodine, 147
ions, 10, 16, 22, 25, 37, 49, 51, 68, 71, 74, 79, 81, 143, 155
IR spectra, 118, 119
Ireland, 1, 27, 28, 33
irradiation, 11, 13, 24, 61, 81
isolation, ix, 133, 134
isotherms, 90, 92, 93, 94, 95, 99, 100, 110, 114, 116, 117

J

Japan, 85

K

kinetics, 54, 78, 151
KOH, 137, 138, 139, 140, 141, 152, 155

L

laboratory studies, 21
lamella, 86
Langmuir–Blodgett (LB) film, 87, 104, 111
Langmuir–Blodgett (LB) technique, ix, 85, 86, 88, 105
lattice parameters, 142
layered double hydroxides, 168
layering, 124
LC-MS, 33
lifetime, 69
light, 13, 14, 24, 30, 31, 32, 53, 61, 83, 87, 106, 109, 121, 122, 123
light scattering, 83
liquid chromatography, 82
liquid phase, 20, 26
lithium, 125, 162, 169
lithium ion batteries, 162
Luo, 33, 70, 77

M

macromolecules, 39
magnetic properties, 167
magnitude, 14, 61
majority, 23
manipulation, 3, 26, 135
mass, 12, 19, 22, 37, 43, 44, 62, 148, 155
mass spectrometry, 43
materials, vii, ix, 9, 21, 24, 25, 42, 51, 53, 54, 59, 61, 63, 64, 65, 66, 79, 86, 87, 110, 133, 134, 135, 136, 140, 148, 151, 153, 154, 155, 159, 160, 161, 162, 166, 169
materials science, 87

matrix, 13, 17, 23, 25, 27, 28, 40, 45, 52, 59, 64, 144
measurements, 21, 91, 93, 110, 113, 123
mechanical properties, 9, 148
media, 150, 151, 154, 168
membrane separation processes, 4, 16
membrane separations, vii, 1, 2, 3
membranes, vii, viii, 2, 4, 5, 7, 8, 9, 10, 12, 13, 16, 17, 18, 19, 21, 22, 23, 24, 25, 26, 27, 28, 29, 30, 31, 32, 34, 35, 36, 37, 38, 39, 40, 86
metal carbides, 135
metal ions, 47
metal nanoparticles, 53
metal oxides, 134, 135, 150, 155, 159
metals, 87, 150
methanol, vii, 1, 25, 26, 40, 59, 62, 79, 91, 143, 147, 151
methyl methacrylate, 126
microfiltration, vii, 1, 3, 13, 27, 30
microorganisms, 14
microRNA, 48, 49, 53
microscope, 93, 111, 123
migration, 24
mixing, 10, 13, 26, 91, 151
models, 19, 20, 119, 120, 123, 146
modulus, 134
molecular orbital, 145, 147
molecular orientation, 124
molecular oxygen, 156, 157
molecular weight distribution, 91, 110, 115
molecules, vii, 1, 2, 4, 16, 17, 18, 20, 22, 23, 24, 26, 30, 34, 46, 51, 61, 83, 102, 105, 144, 146, 151
molybdenum, 44, 45, 68, 70, 71, 72, 73, 74, 75, 76, 78, 79, 80, 81, 82, 160
monolayer, ix, 2, 9, 29, 51, 63, 66, 78, 80, 86, 90, 93, 95, 96, 97, 98, 99, 100, 101, 104, 106, 108, 109, 110, 113, 114, 115, 116, 117, 123, 124, 148, 163, 164, 166
monomers, 90, 109
morphogenesis, 105
morphology, 25, 42, 90, 93, 95, 96, 98, 99, 100, 105, 110, 142
Moses, 74
multivariate analysis, 83
multiwalled carbon nanotubes, 55

N

Na^+, 92
Na2SO4, 9, 11
NaCl, 9, 10
nanocomposite membranes, 5, 26, 27
nanocomposites, 3, 29, 54, 57, 59, 67, 76
nanocrystals, 55, 157, 164
nanofibers, 106
nanofiltration, vii, 1, 3, 7, 14, 16, 24, 27, 28, 30, 31, 33, 34, 35, 36, 37, 38, 39, 40
nanolayers, 168
nanomaterials, 3, 29, 43, 44, 47, 49, 55, 64, 65, 66, 67, 80, 150, 168
nanometer, 18, 87, 105
nanometer scale, 18
nanoparticles, 4, 13, 44, 46, 47, 51, 53, 56, 61, 63, 67, 68, 72, 73, 74, 76, 77, 79, 82, 83, 86, 106, 108, 113, 119, 124, 151, 155, 164, 168
nanoribbons, 45, 68, 149, 150, 157, 166
nanorods, 151, 163
nanostructures, 8, 72, 87
nanotube, 57, 69, 75, 167
national security, 43
neutral, 17, 18, 20, 22, 35, 37
next generation, 61
nickel, 154, 159, 164, 168
NIR, 81
nitrides, 135
nitrite, 55, 74
nitrogen, 81, 141, 143, 149, 151, 156, 161, 162, 164, 165, 166, 167, 169
NMR, 90, 109
Nobel Prize, 42
noble metals, 61, 134, 135, 150, 154
non-polar, 16, 20
nucleic acid, 47, 49

N-vinyl carbazole, 90

O

operations, vii, 1, 2, 3, 23, 24
opportunities, ix, 133, 159
optical properties, 124
optoelectronics, 65
organic compounds, 13, 15, 17, 24, 33, 34, 36
organic matter, vii, 1, 4, 16, 22, 38
orientation state, 122
osmosis, vii, 1, 3, 7, 29, 30
osmotic pressure, 10, 12
out-of-plane X-ray diffraction (XRD), ix, 85, 87, 89, 96, 97, 98, 101, 102, 103, 104, 105, 106, 111, 112
overlap, 46, 120, 122
ox, 151
oxidation, 23, 28, 39, 44, 46, 50, 54, 154, 156, 157, 163, 164, 168
oxidative stress, 33
oxide electrodes, 60
oxide nanoparticles, 44
oxygen, vii, ix, 3, 8, 25, 79, 133, 134, 135, 137, 142, 144, 145, 146, 147, 149, 151, 153, 154, 156, 157, 158, 159, 161, 162, 163, 164, 165, 166, 167, 168, 169
oxygen evolution reaction, vii, ix, 133, 134, 153, 159, 169
oxygen plasma, 8
ozonation, 15, 39
ozone, 15, 23, 24, 39

P

palladium, 76
parallel, 35, 52, 64, 122
partition, 17
pathway, 135, 136, 137, 142,143, 145, 147, 149, 150, 153, 158
peptides, 49, 63
permeability, viii, 2, 3, 8, 12, 26, 36
permeation, 2, 9, 11, 25, 27, 29, 110
permission, 42, 44, 45, 48, 50, 52, 56, 58, 60
peroxide, 23, 145, 156
pervaporation, vii, 1, 3, 11, 12, 31
PES, 8, 11, 19, 37
pH, 8, 17, 18, 20, 24, 30, 40, 61, 64, 143
pharmaceuticals, 2, 4, 14, 15, 16, 18, 20, 22, 24, 33, 34, 35, 36, 38, 39, 40
phase boundaries, 150
phase diagram, 116
phase inversion, 5, 11
phenol, 3, 51, 83
phenylalanine, 36
phosphorus, 142, 145, 162, 164, 167
phosphorylation, 49
photocatalysis, 23, 30
photocatalyst, 13, 39
photodetector, 79
photodynamic therapy, 70
photoluminescence, 71, 81
photonics, 65
physical characteristics, 21
physical properties, 37
physicochemical characteristics, 12
physicochemical properties, viii, 41, 42, 43, 63
physics, viii, 4, 41, 62, 78, 87
piezoelectricity, 63, 79, 80
platform, 48, 49, 53, 55, 59, 61, 65, 70, 71, 72, 73, 76, 77, 81, 135
platinum, 55, 163
PMMA, 10
PNA, 49, 70
polar, 15, 16, 17, 23
polarity, 17, 145
polarization, 21, 22, 63, 118, 142, 143, 148
pollutants, 14, 15, 31, 34, 36, 39
pollution, 14, 32
polyamides, viii, 85, 87, 89, 90, 92, 93, 94, 102, 105, 106
polycondensation, 92
polymer, vii, viii, 5, 8, 10, 22, 26, 63, 73, 85, 86, 87, 89, 90, 95, 98, 99, 101,

105, 106, 107, 108, 109, 110, 111, 113, 115, 116, 118, 119, 122, 123, 124
polymer chain, 95, 99, 101, 117, 123
polymer nanosphere multilayered organization, vii, viii, 85, 89, 101, 106, 107, 111, 122, 123, 124
polymer solutions, 5
polymer synthesis, 115
polymeric membranes, viii, 2, 25, 26, 39
polymerization, 8, 9, 10, 90, 109
polymers, 3, 86, 87, 88, 90, 91, 97, 105, 106, 108, 110, 115
polynucleotide kinase, 49, 71
polysaccharides, 21
polystyrene, 97, 106, 110
polyvinylidene fluoride, 63
population, 145
porosity, 3
portability, 59
precipitation, 5
preparation, 3, 8, 10, 26, 54, 169
principles, 44, 62, 78, 148, 161, 167, 168, 169
prior knowledge, 4
probe, 47, 49, 51, 57, 59, 62, 82, 93, 111
probe station, 93, 111
process control, vii, 1
project, 27
prostate specific antigen, 48, 61, 70
protection, 9
proteins, vii, 1, 3, 5, 47, 61, 77, 86
proton exchange, 25, 26, 40
PTFE, 12
pure water, 7
purification, 3, 31, 90, 109
PVP, 127
pyrolytic graphite, 136

Q

Q–e scheme, 91, 110
quality control, 82
quantification, 64
quantum confinement, 134
quantum dot, 47, 50, 55, 59, 69, 71, 74, 81, 82, 83
quantum dots, 47, 50, 59, 69, 71, 74, 81, 82, 83
quartz, 62, 64, 111
Quartz, 79
quaternary ammonium, 92

R

radiation, 93, 111
reaction mechanism, 136, 153, 156, 161
reaction rate, 46
reactions, vii, ix, 44, 49, 71, 92, 133, 134, 135, 137, 159, 163, 164, 166, 168, 169
reactive sites, 3
reactivity, 149
recognition, 63, 79
recombination, 13
recovery, 19, 47, 61
recycling, 22, 59
refractive index, 87
refractive indices, 87, 90
regulatory framework, 15
rejection, vii, 1, 4, 8, 9, 10, 12, 16, 17, 18, 19, 20, 22, 24, 28, 29, 30, 34, 35, 36, 37, 38
relaxation, ix, 86, 109, 112, 113, 114, 115, 124
relaxation process, 124
renewable energy, 39
repellent, 107
repulsion, 6, 10, 18, 23
requirements, 7, 10, 12, 15, 16
researchers, vii, 1, 44, 46, 49, 51, 53, 54, 59
residues, 8, 39
resistance, 8, 9, 11, 21, 35, 87
resolution, 46, 64, 93, 102
resorcinol, 54, 72
response, 57, 61, 141
restoration, 47
reverse osmosis, vii, 1, 3, 7, 11, 16, 29, 35, 36, 38, 39

riboflavin, 56, 58, 75, 81
rings, ix, 86, 98, 99, 109, 122, 124, 144
RNA, 49, 53
rods, 106
room temperature, 77, 78, 109, 134, 165
roughness, 3, 8, 9, 10, 11, 16, 21, 26, 87
Royal Society, 44, 48, 52, 58

S

salinity, 12
salt concentration, 9
salts, 7, 8, 11, 12, 20, 22
scaling, 22, 35, 38
scarcity, 7, 134
scope, 142
SDS, 68
selectivity, vii, 1, 3, 4, 26, 46, 53, 55, 57, 59, 63, 68, 81, 87, 89, 150
self-assembly, 160
semiconductor(s), 149, 160
sensing, viii, 41, 43, 44, 45, 48, 49, 50, 53, 54, 56, 57, 58, 60, 61, 63, 64, 65, 67, 68, 71, 72, 73, 74, 76, 77, 78, 79, 80, 81
sensitivity, 45, 47, 51, 53, 57, 59, 61, 63, 68, 81
sensor(s), vii, viii, 41, 42, 43, 44, 46, 47, 49, 48, 50, 51, 52, 53, 54, 57, 58, 59, 60, 62, 63, 64, 67, 68, 69, 70, 73, 75, 77, 78, 79, 80, 81, 82
septic tank, 14
serum albumin, 11, 21, 81, 82
sewage, 14, 33
Si3N4, 62
side chain, viii, 85, 87, 90, 93, 109, 110, 119, 123
signal-to-noise ratio, 47
silane, 61
silica, 21, 165
silicon, 165
silver, 8, 29, 49, 71, 82, 83
single-particle layer, viii, 85, 87, 88, 89, 92, 93, 94, 95, 96, 97, 99, 100, 101, 102, 103, 104, 105, 110, 111, 112, 113, 115, 116, 117, 124
SiO2, 62, 82
SO42, 11
sodium, 10, 46, 50, 51, 163
solar cells, 165
sol-gel, 13, 25
solid state, 109
solubility, 17
solution, 3, 5, 10, 12, 18, 23, 24, 26, 44, 60, 64, 77, 90, 91, 92, 109, 122, 135, 136, 150, 151, 153
sorption, 14, 35, 36, 146
speciation, 18
species, 20, 43, 144, 146, 151, 155, 156
specific surface, 143
spectroscopy, 59, 82, 90, 109, 148
speculation, 113
spin, 6, 26, 142, 145, 148, 149, 165
stability, 3, 6, 8, 10, 12, 25, 44, 49, 141, 143, 150, 154
stabilization, 120
stained glass, 107
state, 17, 46, 109, 113, 116, 121, 134, 135, 141, 143, 148, 156, 159, 165, 168
storage, viii, 41, 42, 43, 65, 66, 134, 159, 160, 162
structural formation, 95
structure, vii, viii, ix, 2, 4, 5, 11, 21, 25, 42, 43, 47, 53, 85, 87, 89, 90, 91, 92, 97, 98, 101, 102, 104, 105, 108, 109, 111, 115, 118, 120, 123, 124, 133, 135, 142, 143, 145, 146, 153, 154, 157, 158, 161, 162, 164, 165, 166
styrene, 31
substitution, 157
substrate, 30, 44, 46, 62, 65, 92, 93, 112, 113, 115, 116, 150
sulfur, 62, 81, 142, 145, 162, 165, 167
Sun, v, 27, 28, 29, 30, 32, 39, 65, 72, 73, 74, 83, 133, 143, 151, 160, 161, 163
superlattice, 164
surface area, 42, 134, 141, 142, 150, 154
surface energy, 30

surface layer, 6
surface modification, 8, 32
surface properties, 35, 87, 159
surface structure, 159
surfactants, 10, 44
synergistic effect, 8, 21, 54, 157
synthesis, 12, 28, 39, 55, 59, 63, 65, 68, 73, 79, 81, 145, 160, 161, 162, 164, 165, 166, 167, 168

T

tacticity, 90, 109
target, 10, 47, 48, 49, 57, 59, 62
techniques, ix, 35, 53, 57, 86, 123, 148, 159
technology(ies), vii, 1, 3, 15, 34, 115, 123, 134, 160
TEM, 97, 144
temperature, 8, 16, 26, 93, 95, 96, 100, 109, 116, 124
testing, 24, 39
therapeutics, 67
thermal stability, 93
thermal treatment, 146, 151
thrombin, 48
titania, 9, 29
titanium, 32, 59, 165
titanium dioxide, 13, 32, 59
TNF, 62
TNF-α, 62
toluene, 61, 91
topology, 112, 113, 115
trade-off, vii, 1
transduction, 43
transformation, 14, 15, 32, 33, 134
transformation product, 14, 32, 33
transistor, 43, 59, 60, 77, 78, 79
transition metal, vii, viii, 41, 42, 43, 66, 67, 70, 77, 80, 134, 155, 159, 160
transport, vii, 1, 2, 8, 12, 18, 20, 22, 25, 26, 35, 37, 62, 156, 165
treatment, 4, 14, 15, 16, 24, 27, 29, 30, 31, 32, 33, 34, 35, 36, 38, 39, 104, 142, 152
trifluoroacetic acid, 92
triphenylphosphine, 145
tryptophan, 36
tumor necrosis factor, 62
tungsten, 44, 75, 76, 82

U

uniform, viii, 85, 96, 101, 102, 105, 106, 124
uric acid, 54, 56, 73, 74
UV irradiation, 11
UV spectrum, 118

V

vacancies, 154, 159, 166
vacuum, 6, 10, 12, 13, 26, 87
valence, 142, 143, 157
vanadium, 75, 76, 164
vapor, 76, 78
variations, 19
VDF, 14, 32
velocity, 16, 19, 24

W

waste, 12, 33
wastewater, 4, 14, 15, 16, 21, 33, 34, 36, 38, 39
water, vii, viii, 1, 2, 4, 5, 7, 8, 9, 11, 12, 14, 15, 16, 17, 18, 19, 21, 22, 23, 24, 25, 26, 27, 28, 29, 30, 31, 32, 33, 34, 36, 39, 40, 52, 58, 66, 70, 71, 81, 83, 85, 86, 88, 90, 91, 92, 93, 96, 97, 98, 100, 101, 102, 103, 104, 105, 106, 108, 109, 110, 111,113, 115, 117, 118, 124, 134, 135, 137, 145, 149, 153, 154, 156, 157, 160, 163, 164, 168
water permeability, 3
water purification, vii, 1, 26, 28, 29, 34
water quality, 15
wave number, 87, 89

weak interaction, 109
wealth, viii, 41
weight ratio, 156
wettability, 87
wide band gap, 147, 149
workers, 61, 143, 145, 146, 149, 151, 152, 154, 158

X

X-ray diffraction (XRD), ix, 85, 87, 89, 92, 93, 96, 97, 98, 101, 102, 103, 104, 105, 106, 109, 110, 111, 112, 119, 120, 123

Y

yield, 46, 70, 137, 146, 154, 155

Z

zirconia, 90